Introduction to Synthetic Array and Imaging Radars

Introduction to Synthetic Array and Imaging Radars

S.A.Hovanessian
Hughes Aircraft Company and the University of California, Los Angeles

Book Design
Jane Carey

Copy Editing
Edward Matheny

Typesetting
Linda Grant

ARTECH HOUSE, INC.
610 Washington St., Dedham, MA

Printed and bound in the United States of America.

Standard Book Number 0-89006-082-7

Library of Congress catalog card number: 79-27922

This book is dedicated to

Mary M. Hovanessian

for her continuous support and encouragement.

Table of Contents

LIST OF FIGURES X

LIST OF TABLES XIV

PREFACE XV

INTRODUCTION XVII

1

REAL ARRAY IMAGING RADARS 1

1.1 Resolution Considerations 3
1.1.1 Resolution in the Y-Direction 6
1.2 Discussion of Imaging Radars 10

2

SYNTHETIC ARRAY RADARS (SAR) 13

2.1 Phase and Frequency Relationships 13
2.2 Filter Design Principles 16
2.3 Resolution in the Y-Direction 17
2.3.1 Example 21
2.4 Maximum Resolution 22
2.5 Focused and Unfocused Arrays 23
2.6 Summary and Extension of Equations 27

3

SIGNAL PROCESSING METHODS 31

3.1 Holography and Photographic Zone Plates 31
3.2 Optical Signal Processing 34
3.3 Electronic Signal Processing 38
3.4 Electronic Signal Processing Considerations 43
3.4.1 PRF Selection 45

3.5 Power Return Considerations 48
3.5.1 Atmospheric Absorption 52
3.5.2 Average Power Requirements 60
3.5.3 Power Return with Pulse Compression 61

4

IMPLEMENTATION AND APPLICATION OF SAR 65

4.1 Motion Compensation 66
4.2 Storage Media 67
4.3 Complex Filtering 67
4.4 Transmitter-Receiver Stability 68
4.5 Examples of Synthetic Array Radar Imagery 68
4.5.1 Airborne SAR Imagery 71
4.6 Discussion and Application of Synthetic Array Radars 74
4.6.1 State-of-the-Art SAR Applications 75
4.6.2 SAR Technology 76
4.6.3 The Future of SAR 77

5

PULSE COMPRESSION TECHNIQUES AND SAR MECHANIZATION 79

5.1 Mathematical Derivations 82
5.1.1 Convolution Integral 83
5.1.2 Matched Filter Analysis 84
5.2 Range/Range Rate Ambiguity 87
5.3 Digital Pulse Compression 90
5.4 Mechanization Block Diagram 95
5.4.1 Digital Mechanization 96
5.5 Fourier Analysis 96
5.6 Quadrature Components 100
5.7 Digital Mechanization of the Fourier Transform 102

5.8 Digital Signal Processing Block Diagram 104
5.9 Memory and Speed Requirements 104

6

TWO-DIMENSIONAL CORRELATION – DESIGN EXAMPLES 109

6.1 Stretch Radars 114
6.2 A Design Example 114
6.3 Real Time SAR Image Processing Example 122
6.3.1 Radar System Design Parameters 123
6.3.2 Signal Processing Parameters 125
6.3.3 Power Requirements 129
6.3.4 Signal Processing Mechanization 131
6.4 SAR Equations and Parameters
References 147

INDEX 148

List of Figures

1-1	Ground Mapping Equipment Onboard an Aircraft	1
1-2	Geometry and Signal Processing of a Real Array Imaging Radar	2
1-3	Resolution in the X-Direction	5
1-4	Circular and Rectangular Antennas with their Gain Patterns	6
1-5	Typical SIN X/X (Solid Lines) and Cosine Weighted Antenna Pattern (Dashed Lines)	7
1-6	Antenna Gain Versus Antenna Diameter with Transmit Frequency as a Parameter	9
1-7	Resolution of Ground Points Near the Mapping Radar	10
1-8	Equidistance Points ABC Show Up in the Same Resolution Cell	11
2-1	Transmitted and Received Signals from a Target	14
2-2	Frequency Spectrum of Limited Duration Sinewave	17
2-3	Resolution Element D/2 in the Y-Direction	18
2-4	Actual and Theoretical Y-Direction Resolution in Synthetic Array Radars	20
2-5	Conventional and Synthetic Array (Doppler) Ground Maps Obtained by an Airborne Radar	21
2-6	Range to a Point on the Ground as the Aircraft Flies with a Velocity V	23
2-7	Quadratic Phase Angle from the Target with Appropriate Adjustments and Generation of Zone Plate	26
2-8	Resolutions in the Y-Direction as a Function of Range for the Cases Focused, Unfocused and Conventional Radars	27
2-9	Flight Geometry for a Squint Angle of θ	29

3-1 Hologram Construction on a Photographic Plate 33

3-2 Reconstruction of the Original Objects from the Hologram 33

3-3 Production of Photographic Zone Plates by Interfering Signals 34

3-4 Construction/Reconstruction of Photographic Zone Plates Using Plane and Spherical Waves 35

3-5 Holographic Image Produced from Recorded Zone Plate Data 36

3-6 Optical Data Processing from Synthetic Array Recorded Data 37

3-7 Phase and Frequency Excursions of a Single Point on the Ground as a Function of Time 38

3-8 Aircraft Flight and Frequency History of Points on the Ground 40

3-9 Range Gates Followed by Doppler Filters for Electronic Data Processing 42

3-10 Maximum and Minimum Range-to-Ground for the 3 dB Antenna Beamwidth θ_{BW} 43

3-11 Ground Coverage in the Y-Direction for a 3 dB Beamwidth θ_{BW} 46

3-12 Transmitted Pulses and Spectral Diagrams of PRF Lines 47

3-13 Backscattering Coefficients for Oceans and Other Surfaces 50

3-14 Backscattering Coefficient at Low Angle of Incidence (5.5 Degrees) and Frequency of 16.4 GHz 53

3-15 Absorption in the Atmosphere Caused by Water Vapor (Radar at Sea Level) 54

3-16 Absorption in the Atmosphere Caused by Oxygen (Radar at Sea Level) 55

3-17 Absorption in Atmosphere by Free Electron and Combined Oxygen and Water Vapor (Radar at Sea Level) 57

3-18	Attenuation of Electromagnetic Energy by Atmospheric Gases at Standard Atmosphere	58
3-19	Rain Attenuation as a Function of Wave Length a) Drizzle – 1/4 mm/hr, b) Light Rain – 1 mm/hr, c) Moderate Rain – 4 mm/hr, d) Heavy Rain – 16 mm/hr	59
3-20	Attenuation of Electromagnetic Energy Due to Fog – Light Fog 0.032 g/m³, Heavy Fog 2.3 g/m³	59
4-1	Synthetic Array Radar Picture of the Coast of Mexico Obtained via Satellite	70
4-2	Synthetic Array Radar Picture of the Florida Coast Obtained via Satellite	70
4-3	Synthetic Array Radar Imagery Obtained from Airborne SAR	72
5-1	Transmitted and Received Pulses	79
5-2	Transmitted Waveform of a Linear FM Pulse	80
5-3	Received Waveform of the FM Pulse and Subsequent Pulse Compression	81
5-4	Amplitude-Time Characteristics of the Compressed Pulse	81
5-5	Sidelobe Suppression of Linear FM Compressed Pulse at the Matched Filter Output	83
5-6	Amplitude-Frequency-Time Relationship of a Pulse Compression Signal	87
5-7	Time in Pulse Compression Filter Before and After Peak Value of Non-Moving Target Returns	89
5-8	Waveforms for a Barker Code of Length 13	90
5-9	Barker Decoder of Length 13	91
5-10	Procedure for Calculating the Decoder Output for Barker Code Containing 13 Elements	91
5-11	Barker 13 Decoded	92
5-12	Calculation of the Correlation Function of Frank Code 16 Elements	94

5-13	Frank Code of 16 Elements	94
5-14	SAR Mechanization Block Diagram	95
5-15	Discrete Representation of F(t) and G(ω)	98
5-16	Quadrature Components of Input Signal	101
5-17	Input Signal as a Series of Rotating Phasors	102
5-18	Digital Filter for the Computation of G_n	103
5-19	SAR Mechanization Block Diagram with Digital Signal Processor	105
6-1	Two-Dimensional Synthetic Array Correlator	109
6-2	Returns from Several Transmitted Pulses are Range Gated and Filtered	110
6-3	Doppler Shift Measurement from 20 Returned Pulses	111
6-4	a) Presence of Multiple Frequencies in Time Signal, and b) its Resolution into Amplitude-Frequency	112
6-5	Range or X-Direction Resolution with Pulse Compression	112
6-6	Azimuth or Y-Direction Resolution with Received Frequency Excursion	113
6-7	Compressed Pulse Signal Processing	115
6-8	Aircraft Flight of the Design Example	117
6-9	Spacecraft Geometry of Example Problem	122
6-10	SAR Signal Processor Block Diagram	131
6-11	Range Gates and Frequency Modulated Returns	132
6-12	Range Correlator Consisting of Transverse Filter	133
6-13	Frequency-Time Relationship of Azimuth Returns	134
6-14	Set of 55 Parallel Azimuth Filters for Azimuth Correlation	135
6-15	Filter Data Dumps	136

List of Tables

Table		Page
1-1	RESOLUTION CELL SIZE FOR VARIOUS OBJECTS	3
3-1	RADAR TARGET CROSS SECTIONS	52
4-1	SEASAT-A SPACEBORNE SAR CHARACTERISTICS	69
4-2	DEMONSTRATED SAR APPLICATIONS	75
6-1	RADAR SYSTEM DESIGN PARAMETERS	138
6-2	SIGNAL PROCESSING PARAMETERS	139
6-3	SPACECRAFT SUPPLIED PARAMETERS	140
6-4	SAR EQUATIONS AND PARAMETERS	141

Preface

This book is the outgrowth of the portions of the lecture notes of a 5-day course on synthetic array and imaging radars which the author had the privilege of organizing and, together with his colleagues, presenting at the University of California, Los Angeles (1978). Most of the participants in the course had some radar background and were primarily interested in learning how radars are used to obtain ground maps.

The book contains introductory coverage of synthetic array and imaging radars with all of the related principles and mathematical derivations obtained from basic physical laws. Special care has been taken to use simplest derivations in obtaining needed mathematical relations. In most cases these relations are obtained by using nothing more than simple algebra.

After a brief introduction to mapping radars, a discussion of real array imaging radars is presented in Chapter I. Chapter II gives the basis of synthetic array radars (SAR) together with all of the relevant mathematical relations. Chapter III contains both optical and electronic signal processing methods applied in obtaining synthetic array ground maps. This chapter also includes power return considerations, PRF selection, etc. Chapter IV gives a general discussion of the parameters and conditions involved in obtaining high quality images from synthetic array radars. It includes examples of spaceborne and airborne SAR imagery together with a discussion of probable future SAR developments.

Chapter V contains a complete discussion of pulse compression methods which are commonly applied in synthetic array radars. This chapter also contains a great deal of information regarding digital pulse compression methods and related applicable codes. Mechanization block diagrams together with Fourier transform and spectral analysis methods are given in this chapter. The chapter concludes with a detailed discussion of digital signal processing requirements. The concluding chapter, Chapter VI, gives a discussion of two-dimensional correlators used in obtaining two-dimensional radar maps. This chapter also includes two synthetic array radar design examples. One example is that of a SAR system on-board an airplane; the other is that of a SAR system on-board a spacecraft. A table which includes all of the important SAR equations derived in this book is also given at the end of this chapter.

The author wishes to acknowledge the participation of his colleagues Messrs. J.J. Kovaly, J.M. Swiger and F.C. Williams in presenting the course at UCLA and their subsequent review of the manuscript. Their valuable comments served to improve the contents of this text and enhance the clarity of its presentation.

In addition, the author is grateful to Messrs. Frederick V. Stuhr and Frank T. Barath both of the Jet Propulsion Laboratory, California Institute of Technology, for providing some of the SAR applications and discussion material used in Chapter IV. Mr. E.E. Herman of Hughes Aircraft Company made a number of valuable suggestions and recommendations for which the author expresses his gratitude.

Los Angeles, California
November 1979

S.A. Hovanessian

Introduction

The first aerial images of the earth were crude photographs taken from balloons in the middle of the 19th century. As cameras and photographic films and processing — together with airplanes and space flights — evolved, aerial imagery was employed to study and record earth's surface conditions. Today, nearly all geological and topographic maps are based on images obtained from aircraft or artificial satellites.*

In the 1950's, imaging devices were invented with sensitivities beyond the visible range of wavelengths of 0.4 to 0.8 microns (4×10^{-5} to 8×10^{-5} cm), into the infrared (IR) region of 0.8 to 15 microns. Thus, these devices detected energy that was either sunlight reflected from the terrain or from man-made structures, or was radiated by them as a function of their temperature.

At visible and infrared wavelengths, however, the atmosphere absorbs a significant fraction of radiation, even in the clear environment. In cloudy or rainy weather, the performance of visible and infrared detectors is seriously impaired because of the fact that a great amount of emitted energy is absorbed by the atmosphere. To overcome this dependence on weather and atmospheric conditions, an imaging device was needed which provides its own source. Additionally, the source of illumination of this imaging device needed to be one that would not incur great losses of energy when traveling through the atmosphere.

Radar systems, operating at a wavelength of one to 30 centimeters, fulfill both of the above requirements. Firstly, these systems provide their own source of illumination by transmission of electromagnetic energy, and secondly, at wavelengths of greater than one centimeter the atmospheric absorption of transmitted and reflected energy is minimal. Thus, radar imaging systems can essentially provide an all-weather operation.

Note that radar systems operating at wavelengths in the order of centimeters, compared to the visible region of the order of 10^{-5}

---

*Reference 8.

centimeters, cannot provide the same level of detail image resolution obtainable in the visible region. The radar wavelengths, however, are still short enough to obtain imagery displaying remarkable proximity to actual terrain for both geographical and geological identification purposes.

Synthetic Array and Imaging Radars

Synthetic array and imaging radars are airborne and spaceborne radars which obtain a "photograph" of the ground below through transmission and reception of electromagnetic energy. As is well known, the clarity of a two-dimensional photograph will depend on the resolution obtained in each dimension, i.e., x- and y-directions. The x-direction is considered perpendicular to the flight path, while the y-direction is considered to be along the flight path of the vehicle.

Imaging radars can be divided into two broad categories; real array and synthetic array. Imaging radars, both real and synthetic array, obtain x-direction resolution by controlling the transmitted pulsewidth. Real array radars obtain the y-direction resolution by the antenna beamwidth. Since the antenna beamwidth in each direction is inversely proportional to its length in that direction, the y-direction resolution of real array imaging radars is obtained through the use of long antennas (narrow beamwidth) usually mounted on the belly of the aircraft. Real array imaging radars are also referred to as conventional real array radars or conventional mapping radars.

The desire to use regular size airborne antennas for ground mapping led to the invention of synthetic array radars (SAR)*. These radars use the transmitted pulselength for the x-direction resolution, similar to real array imaging radars. In the y-direction, however, synthetic array radars use the incremental doppler shift of adjacent points on the ground for resolution, rather than the antenna beamwidth, as was the case in real array imaging radars.

---

*Also called synthetic aperture radar. In antenna theory the term array is usually used for antennas in which individual radiating elements are separate rather than part of a continuous radiator. The term aperture defines a class of continuous radiators over a given surface (aperture).

Thus, the reduction of antenna size in going from real to synthetic array radars has been, in effect, replaced by the increased amount of doppler (frequency) signal processing required for the y-direction resolution.

The development of synthetic array radar originated in 1951 with Carl Wiley, who postulated the use of doppler or frequency information in obtaining y-direction resolution. Based on this idea and succeeding developments, the first SAR image was produced by researchers at the University of Michigan in 1958 using an optical processing method.

Precision optical processors and hologram radars were developed and fine resolution strip maps obtained by the mid 1960's. The development of digital computers and digital signal processing methods was utilized by synthetic array radars as an alternative to optical signal processing methods. In the early 1970's, SAR imagery using digital signal processing methods were obtained off-line or *non-real* time. The present trend indicates the utilization of digital signal processing for the production of *real-time* SAR imagery.

It should be noted that a radar may be built for the specific purpose of ground mapping, or ground mapping may be incorporated as one of the several operating modes of a given radar. The additional radar capabilities in these multi-mode radars may include target search and surveillance, target acquisition and tracking, and terrain following and avoidance.

1 Real Array Imaging Radars

Figure 1-1 shows the essential elements of an airborne real array imaging radar. These include radar and navigation equipment, recording and data processing equipment, television and cartographic cameras to provide complementary coverage of the terrain below, and sidelooking antenna and gimbal mechanism. Real array imaging radars obtain y-direction resolution by virtue of the real antenna beam coverage on the ground which is determined by the physical size of the antenna. Finer resolution will require longer antennas, which because of their size can only be mounted lengthwise along the aircraft. This results in an antenna beam looking to the side of the aircraft. For this reason the real array imaging radars are also called side-looking airborne radars.

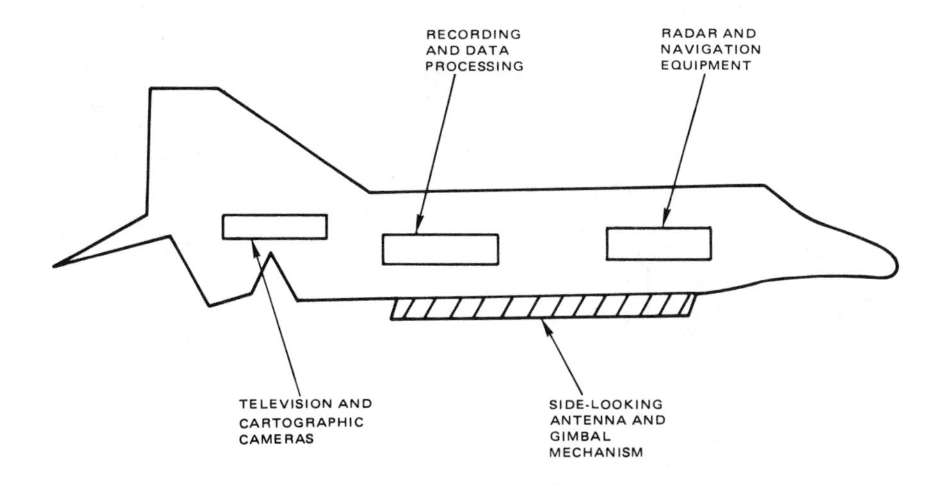

Figure 1-1 Ground Mapping Equipment Onboard an Aircraft

This name, fortunately, also applies to synthetic array radars. As will be seen later, incremental doppler shift of adjacent resolution cells in the y-direction can only be separated if the antenna is looking to one side rather than straight ahead.

Figure 1-2 shows a simplified diagram of geometry and signal processing of a real array imaging radar. The aircraft flies along the track y-direction at velocity V. Its long antenna produces a fan beam illuminating the ground below. Note the narrow width of the antenna beam along y-direction as opposed to the width along the x-direction. As mentioned before, the y-direction resolution is determined by the beamwidth while across the track, the x-direction resolution is determined by the pulse length, i.e., independent of the beamwidth. The transmit/receive function switches the radar between transmission and reception of signals. The received ground returns are displayed on a cathode ray tube and recorded on a photographic film.

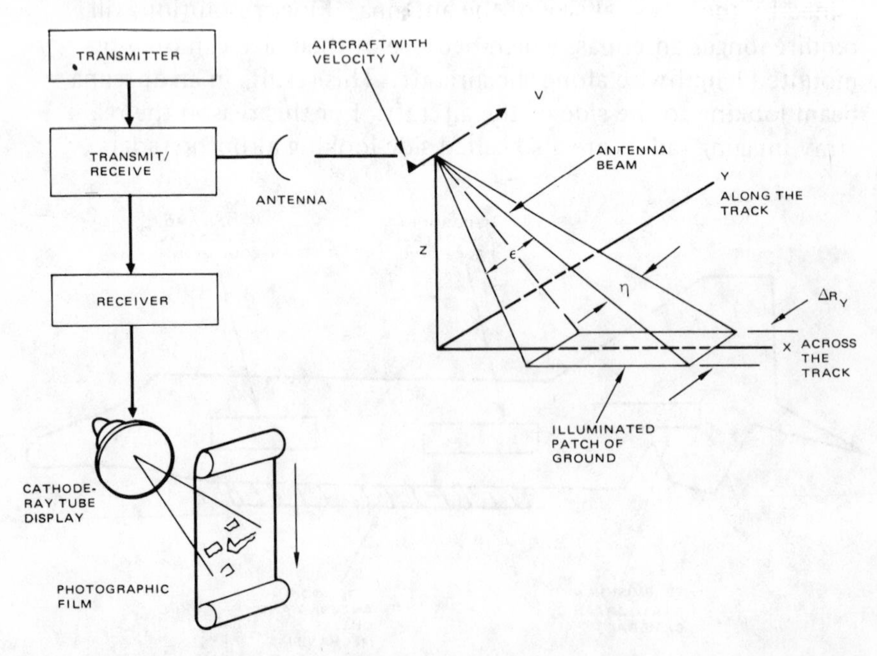

Figure 1-2 Geometry and Signal Processing of a Real Array Imaging Radar

1.1 RESOLUTION CONSIDERATIONS*

Before a discussion of achievable radar mapping resolution is begun, let us examine the required resolution for recognizing outlines, boundaries, and detailed differences of various mapped objects. Table 1-1 gives the required square resolution cell sizes for a variety of objects. From this table it is noted that broad outlines of large areas are recognizable with a 150 m resolution element, while small objects, such as vehicles, require a resolution cell size on the order of meters.

TABLE 1-1 RESOLUTION CELL SIZE FOR VARIOUS OBJECTS

ITEM	SQUARE CELL SIZE, meters
Coastlines, Cities, Mountains	150
Major Highways, Large Airfields	30
City Streets, Large Buildings	15
Vehicles, Houses, Buildings	3

Pulse radars measure range (distance between the radar and the target) by transmitting a pulse and timing the returned pulse from the target as shown in Figure 1-3. This figure shows transmitted pulses of duration τ with an interpulse period of T. The aircraft-ground geometry in the vertical plane is also shown in the figure. Assuming that the range gates are of duration τ, two adjacent points on the ground, A and B, which are ΔR_x apart, can be resolved by their returns occurring in two adjacent range gates as shown. Since electromagnetic energy travels at the speed of light, c, we can write

$$t = \frac{2R}{c} \; ; \; t + \tau = \frac{2(R+\Delta R)}{c} \quad , \qquad (1.1)$$

*For a general text on radars see Reference 5.

where the factor 2 in the above equations signifies round-trip travel of electromagnetic energy to the target. Subtracting equations (1.1) we get

$$\tau = \frac{2\Delta R}{c} \tag{1.2}$$

Also, from Figure 1-3

$$\Delta R_x = \frac{\Delta R}{\cos\alpha}, \tag{1.3}$$

and, finally

$$\Delta R_x = \frac{c\,\tau}{2\cos\alpha}, \tag{1.4}$$

where ΔR_x is the resolution in the x-direction or perpendicular to the track. From equation (1.4) resolution in the x-direction is directly proportional to the pulsewidth τ, i.e., short pulsewidths produce better resolution.*

In matched transmit/receive radars, usually, the product of pulse duration τ and the receiver bandwith W is unity, i.e.,

$$\tau W = 1; \quad W = \frac{1}{\tau} \tag{1.5}$$

Using this in (1.4) we get

$$\Delta R_x = \frac{c}{2W\cos\alpha} \tag{1.6}$$

In the pulse radar of Figure 1-3, the maximum measurable range R_{max} is a function of interpulse period T, and is given by

$$R_{max} = \frac{cT}{2} \tag{1.7}$$

*In most of the derivations given in this book it is assumed that the antenna beam is perpendicular to the flight path. This is done to simplify the mathematical derivations. If this is not the case, an adjustment involving the angle between the flight path and antenna beam can be easily incorporated.

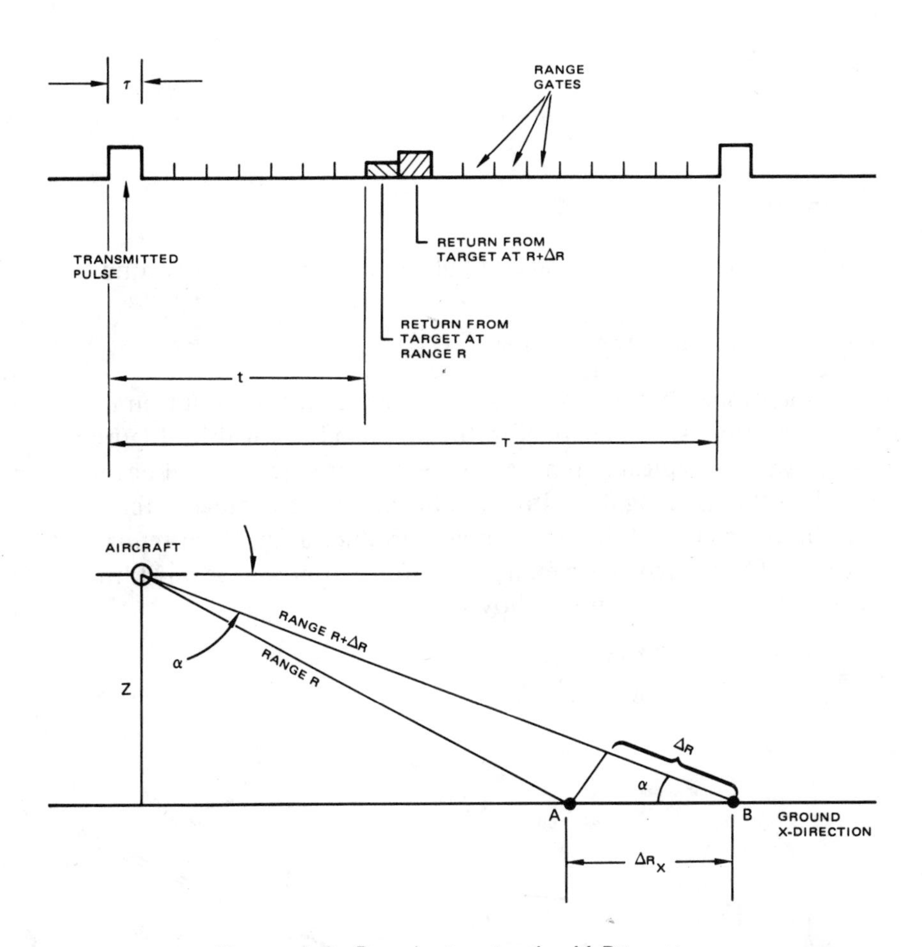

Figure 1-3 Resolution in the X-Direction

Beyond this range, the elapsed time between transmitted pulse and target return pulse will include multiples of interpulse period T, making the measured range ambiguous.

As an example of resolution along the x-direction, consider a radar with a pulse duration of $\tau = 10^{-7}$ seconds with a small look-down angle ($\cos \alpha \cong 1.0$). Using these values in equation (1.4) we get

$$\Delta R_X = \frac{3 \times 10^{10} \times 10^{-7}}{2}$$

$$= 1500 \text{ cm}$$

$$= 15 \text{ meters} \quad ,$$

where 3×10^{10} is the velocity of light in centimeters per second.

1.1.1 Resolution in the Y-Direction

As mentioned before, in real array imaging radars resolution in the y-direction is proportional to the antenna beamwidth. Figure 1-4 shows two typical antennas – one circular with a pencil beam and the other rectangular with a fan beam. A rectangular antenna with uniform current distribution will produce a sin x/x antenna pattern. The approximate expressions for half-power beamwidth of this antenna are given as follows

$$\eta = \frac{51\lambda}{\ell_1} \; ; \quad \epsilon = \frac{51\lambda}{\ell_2} \quad , \tag{1.8}$$

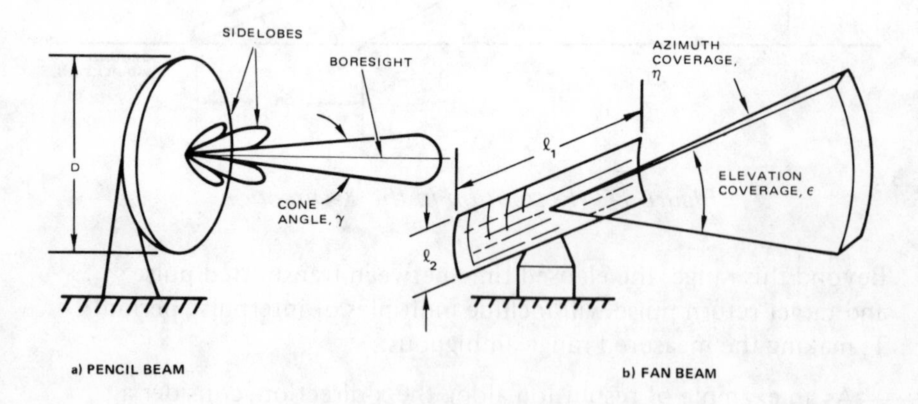

Figure 1-4 Circular and Rectangular Antennas with Their Gain Patterns

where η and ϵ are half-power beamwidths in degrees, λ is the wavelength and ℓ_1 and ℓ_2 are the sides of the antenna as shown in Figure 1-4. A circular antenna with uniform current distribution will produce a radiation field intensity involving first order Bessel functions. The half-power beamwidth in this case will be

$$\gamma = \frac{58.5\lambda}{D} \text{ (degrees)} \tag{1.9}$$

$$\cong \frac{\lambda}{D} \text{ (radians)} \quad ,$$

where γ is the half-power beamwidth and D is the diameter of the antenna as shown in Figure 1-4. In subsequent discussions the value of γ will be approximated by $\gamma = \lambda/D$ where γ is given in radians.

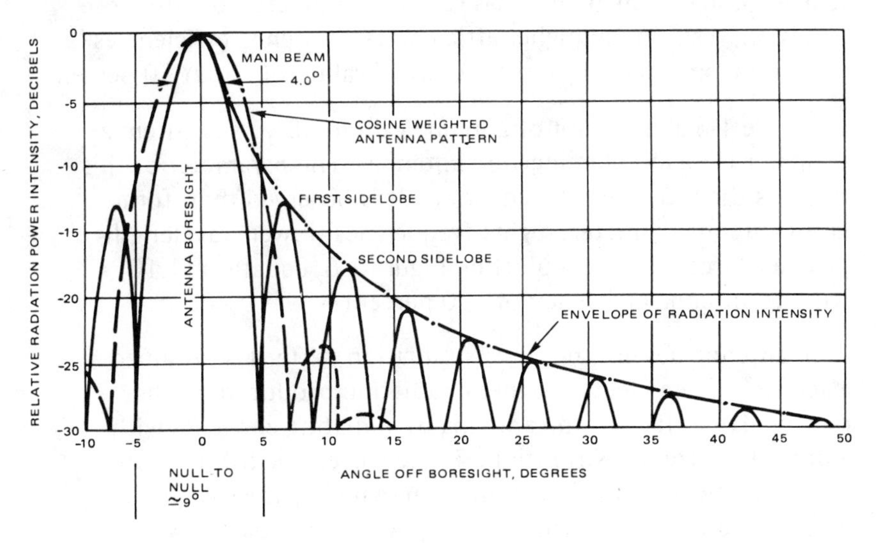

Figure 1-5 Typical Sin X/X (Solid Lines) and Cosine Weighted Antenna Pattern (Dashed Lines)

A typical complete sin x/x antenna pattern for an antenna aperture of 50 inches and a transmitted wavelength of $\lambda = 10$ cm is shown in Figure 1-5. The half power beamwidth of this antenna from equation (1.8) will be

$$\gamma = \frac{51 \times 10}{50 \times 2.54} = 4.0 \text{ degrees} \quad ,$$

as shown in the figure. The first antenna sidelobe is 13.2 dB down from the mainlobe as shown.

The maximum antenna gain, occurring at zero boresight, is given by the relation

$$G = \frac{4\pi A}{\lambda^2} \quad , \tag{1.10}$$

where G is the gain and A is the antenna aperture area. The antenna gain of equation (1.10) assumes a 100 percent antenna efficiency. Antenna efficiency is a function of feed structure and manufacturing surface tolerances, and varies greatly among antennas. Usually small antennas can be manufactured with close precision, resulting in higher efficiencies. Antenna efficiencies can vary from over 90 percent to lower values of about 50 percent.

Figure 1-6 gives a plot of antenna gain in dB versus antenna diameter for a circular aperture antenna. The parameter of this figure is the transmitted frequency. It is observed that, for a given antenna diameter, higher frequencies (lower wavelengths) produce larger gains. In plotting Figure 1-6, equation (1.10) – with an antenna efficiency of 100 percent – was used.

In antenna design, the sin x/x radiation pattern is obtained effectively from electromagnetic radiation produced on the rectangular surface of the antenna by radiating elements with a uniform current distribution. If the current distribution of the radiating elements is not uniform, which is the case in weighted antenna patterns, typically wider mainbeams with lower sidelobes will result. Figure 1-5 also shows a cosine weighted pattern. In a number of cases weighted patterns may be more desirable as they reduce target returns from antenna sidelobes. Note that the sidelobes of a circular antenna with uniform current distribution are lower than that of rectangular antennas. In the case of circular antenna the first sidelobe, for example, is 17.6 dB down from the mainlobe as compared to 13.2 dB for a rectangular aperture.

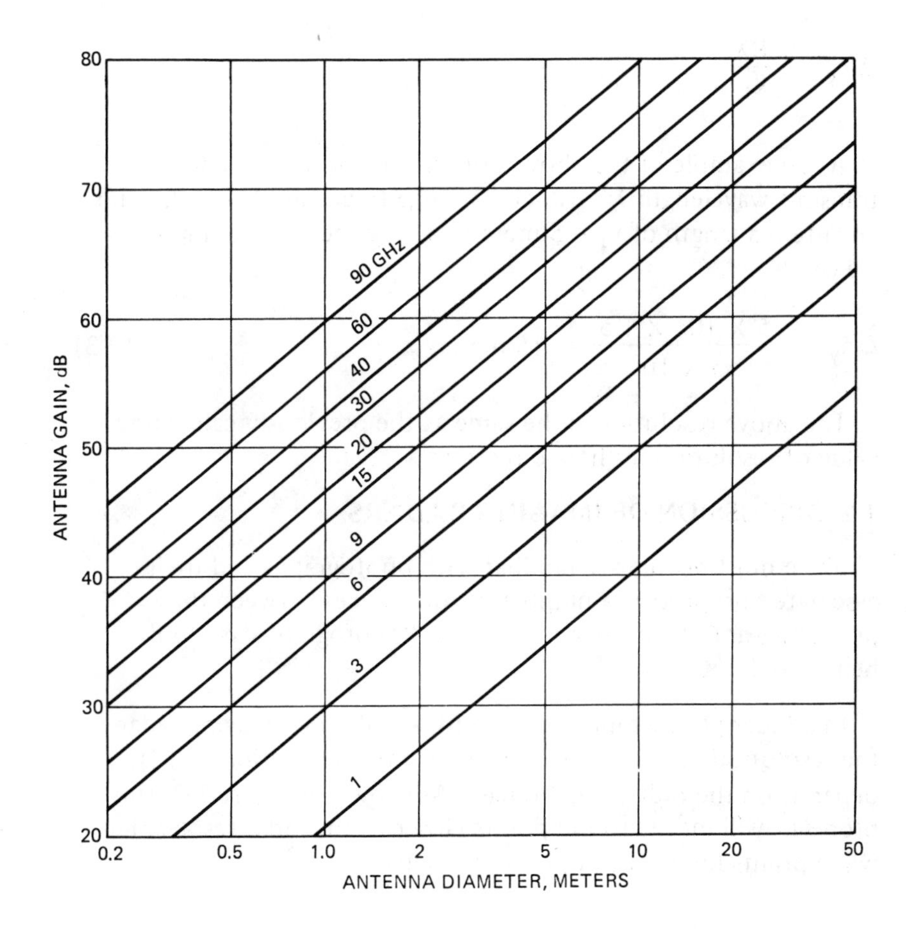

Figure 1-6 Antenna Gain Versus Antenna Diameter with Transmit Frequency as a Parameter

Now it becomes a simple matter to determine the resolution in the y-direction. From Figure 1-2 we note that the resolution in the y-direction is

$$\Delta R_y = R\eta \quad , \tag{1.11}$$

where ΔR_y is the resolution in the y-direction, R is the range to the resolution cell and η is antenna azimuth beamwidth. Approximating the value of η by λ/ℓ_1 we get

$$\Delta R_y = \frac{R\lambda}{\ell_1} \tag{1.12}$$

As an example of the above resolution, consider a radar with a transmit wavelength $\lambda = 1.5$ cm, a range to ground $R = 5$ km and an antenna length of $\ell_1 = 5$ meters. Using these values in (1.12) we get

$$\Delta R_y = \frac{5 \times 10^5 \times 1.5}{5 \times 10^2} = 15 \text{ meters} \tag{1.13}$$

The above resolution is the same as the previously calculated value of resolution in the x-direction.

1.2 DISCUSSION OF IMAGING RADARS

Since most people are familiar with photographs and their associated properties, a number of differences between these properties and the properties of radar "photographs" should be pointed out.

In a radar photograph obtained by a real array imaging radar the resolution in the y-direction as seen from equation (1.12), depends on the radar target range. As shown in Figure 1-7, this property will make ground points closer to the radar resolvable while points farther away will not resolve.

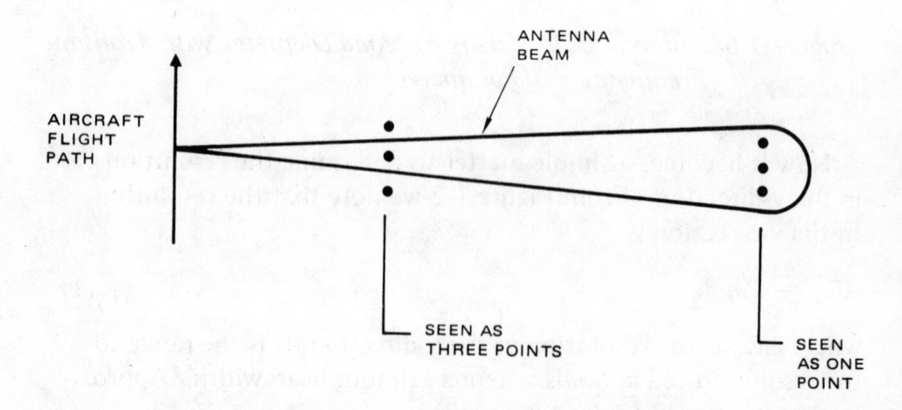

Figure 1-7 Resolution of Ground Points Near the Mapping Radar

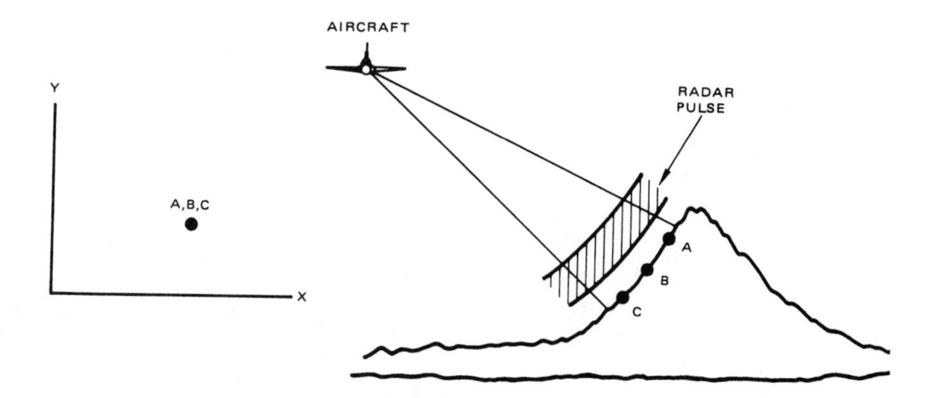

Figure 1-8 Equidistance Points ABC Show Up in the Same Resolution Cell

Since in both real and synthetic array imaging radars the resolution in the x-direction depends on the time of the arrival of the target return signal, equidistance points from the radar will all converge to the same range gate. Figure 1-8 shows three ground points at various elevations which are equidistance from the radar. These points will all show up at the same x-resolution cell. To an observer at the position of the aircraft (or in a photograph) these points will, of course, show up at three distinct locations.

The first large scale project (1968) involving a real array imaging radar was the mapping of the province of Darien, which connects Panama and South America. In 1968 Westinghouse Electric Corporation, in cooperation with Raytheon Company, employed a conventional side-looking radar system, such as described in Figure 1-1, to make images for a mosaic of an area 20,000 km^2. The area had not been mapped previously in its entirety because of an always present cloud cover.*

*Reference 8.

2 Synthetic Array Radars

In order to obtain ground maps without using large antennas, synthetic array radars were invented. In these radars, resolution in the x-direction is determined by pulse duration as in the case of real array imaging radars. The resolution in the y-direction, however, is determined by using doppler shifts or spectral analysis as opposed to the beamwidth, which was the case in real array imaging radars. Thus, in synthetic array radars the long antenna producing narrow beam, in effect, is replaced by a great deal of spectral signal processing. Using synthetic array signal processing procedures one can obtain a high resolution ground map using standard antennas in aircraft or satellites.

Since, in the case of synthetic array radars, the basis of resolution in the y-direction is frequency shift (doppler), which in turn is the rate of change of phase, we will start the discussion with a review of these topics.

2.1 PHASE AND FREQUENCY RELATIONSHIPS

Consider the transmitted waveform of Figure 2.1

$$V_t = \sin 2\pi f_o t \tag{2.1}$$

Assume this signal is reflected from a moving target with radar-target range R given by the equation

$$R = R_o + \dot{R}\, t \, , \tag{2.2}$$

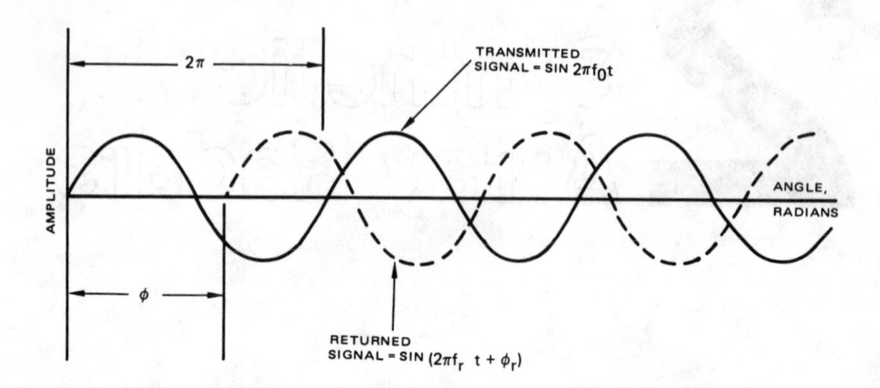

Figure 2-1 Transmitted and Received Signals from a Target

where R_o is the initial range, $\dot{R}$ is target's closing (opening) rate in the direction of R, and t is the time. The returned signal at time t will be the signal which was transmitted Δt seconds ago with

$$\Delta t = \frac{2R}{c} \tag{2.3}$$

Substituting $t = t - \Delta t$ in (2.1) we get the received signal

$$V_r = \sin\left[2\pi f_o \left(t - \frac{2R_o + 2\dot{R}t}{c}\right)\right] \tag{2.4}$$

Grouping terms involving t together we get

$$V_r = \sin\left[t\left(2\pi f_o - \frac{2\dot{R}(2\pi f_o)}{c}\right) - \frac{2(2\pi f_o)R_o}{c}\right] \tag{2.5}$$

Note that in the final analysis of trigonometric functions the coefficient of time t in the argument represents a frequency while the part not involving time represents a phase ϕ. Using this the returned signal V_r of (2.5) will have a frequency of

$$2\pi f_r = 2\pi f_o - \frac{2\dot{R}(2\pi f_o)}{c}\,, \tag{2.6}$$

and a phase

$$\phi_r = -\frac{2(2\pi f_o)R_o}{c} \tag{2.7}$$

Using the value of transmitted frequency $2\pi f_o$ from (2.1) in (2.6) we obtained the difference of transmit and receive frequencies $f_r - f_o$ as follows

$$\begin{aligned} f_d = f_r - f_o &= -\frac{2\dot{R}(f_o)}{c} \\ &= -\frac{2\dot{R}}{c/f_o} \\ &= -\frac{2\dot{R}}{\lambda}, \end{aligned} \tag{2.8}$$

where f_d denotes the doppler shift and c/f_o is the wave length λ. Thus, the return from a moving target will be shifted in frequency directly proportional to radar-target closing rate and inversely proportional to wavelength.

The phase angle of the returned signal ϕ_r as given by (2.7) can be written as follows

$$\begin{aligned} \phi_r &= -(2\pi)\frac{2R_o}{c/f_o} \\ &= -2\pi\frac{2R_o}{\lambda} \end{aligned} \tag{2.9}$$

This equation represents the number of equivalent wavelengths to the round trip distance $2R_o$. The multiplier 2π converts the number of wavelengths to angle in radians. Thus, the phase change is equivalent to a distance. In a similar manner, the doppler shift of equation (2.8) for an incoming target can be interpreted as the number of wavelengths per second that the transmitted signal is "pushed in" due to closing rate $\dot{R}$. The number of wavelengths per second is, of course, frequency.

Equations (2.8) and (2.9) show that the difference between transmit/receive frequencies is proportional to *rate of change* of the radar-target range while the difference in phase of transmit/receive signals is proportional to radar-target range. This is a restatement of the fact that phase is associated with range while the derivative of phase with respect to time (frequency) is associated with range rate.

2.2 FILTER DESIGN PRINCIPLES

In radar systems the frequency shifts discussed above are measured by the spectral analysis of returned signals from the target. These analyses may be carried out either by analog or digital methods. Both methods basically convert a signal represented as a function of time (amplitude-time) to a signal representing frequency contents (amplitude-frequency). In analog systems, transition from time to frequency domain is implemented by tuned filters, while in digital systems this transition is effected by a computer performing Fourier analysis of input time signal. The following is basic to either method.

Figure 2-2 shows the frequency equivalent of a time signal consisting of a sinewave of duration T. It is noted that this time signal can be represented by the "sum" of sine/cosine functions with the amplitude frequency values taken from solid curve of Figure 2-2b, much the same way that any time signal can be represented by a Fourier series. Note the sin x/x behavior of the frequency function and the fact that the null-to-null frequency spread is 2/T. The 3-dB or half power bandwidth occurs at approximately 1/T with the first sidelobe to mainlobe power ratio of -13.2 dB. These characteristics are, of course, similar to the antenna patterns of Figure 1-5. Much the same way as for antennas, various weighting functions can alter the characteristics of the mainlobe and sidelobe of filters as the requirements arise.

From Figure 2-2b the "time constant – bandwidth" relationship of $T = 1/\Delta f$ can also be obtained, where Δf represents the filter bandwidth at 3-dB power points. It should also be noted that the null-to-null bandwidth is $2\Delta f$.

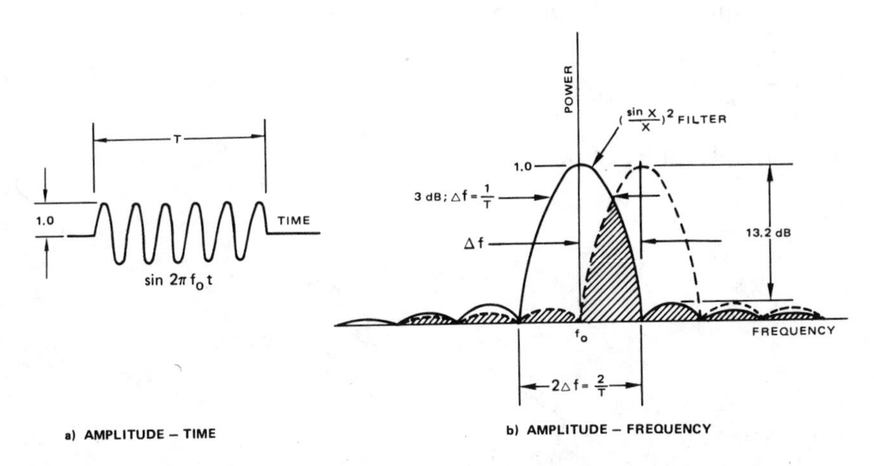

Figure 2-2 Frequency Spectrum of Limited Duration Sinewave

Now, assume that a filter with sin x/x response characteristics is implemented as shown by the solid curve of Figure 2-2b. We desire to obtain the response of this filter to a time function shifted in frequency by the bandwidth Δf, i.e., the time function sin $2\pi(f_o + \Delta f)t$. The frequency spectrum of this signal is shown by the dashed curve of Figure 2-2b. This amplitude-frequency response is the same as that of sin $2\pi f_o t$ except it is shifted by Δf. The filter response to the input sinewave of frequency $f_o + \Delta f$ is represented by the dashed area of the figure where the two curves overlap. Thus, the highest amplitude present will be 3 dB lower in power than that of a sinewave at center frequency f_o. Additionally, energy will be present in filter sidelobes as shown. The presence of energy in the sidelobes, as will be discussed in Chapter IV, affects synthetic array image quality.

2.3 RESOLUTION IN THE Y-DIRECTION

To obtain the resolution in the y-direction consider Figure 2-3. Aircraft flies at a speed of V parallel to the y-axis. We intend to find frequency difference between two ground points A and B which are D/2 apart. From previously given equations

f_B = doppler shift of point B

$$= \frac{2V\cos\theta}{\lambda}$$

f_A = doppler shift of point A

$$= \frac{2V\cos(\theta+\Delta\theta)}{\lambda} \tag{2.10}$$

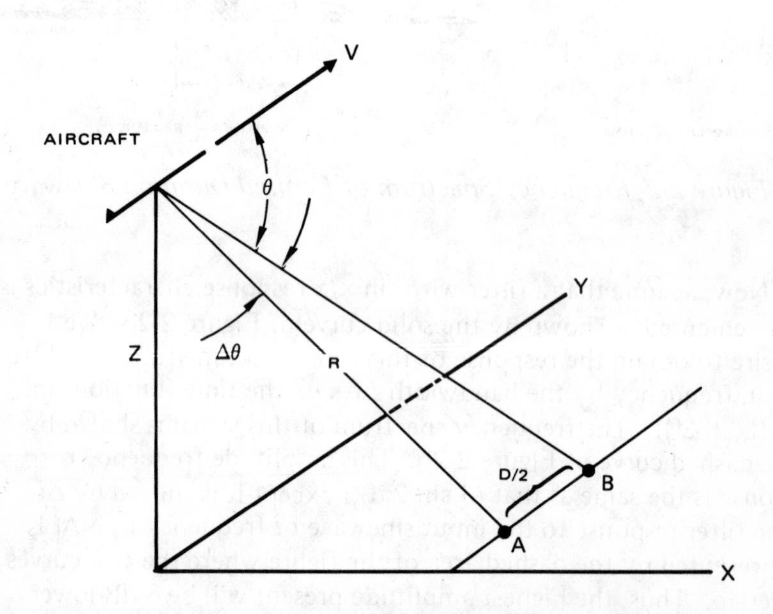

Figure 2-3 Resolution Element D/2 in the Y-Direction

Expanding cos $(\theta + \Delta\theta)$ in terms of sines and cosines and using the small angle relationships cos $\Delta\theta \cong 1$ and sin $\Delta\theta \cong \Delta\theta$, we get

$$\Delta f_i = f_B - f_A = \frac{2V}{\lambda}\Delta\theta \sin\theta \tag{2.11}$$

The value of $\Delta\theta$ from Figure 2-3 can be approximated by

$$\Delta\theta \cong \frac{D}{2R}, \tag{2.12}$$

where R is the range. Using this in (2.11) we get

$$\Delta f_i = \frac{VD}{\lambda R} \sin\theta \tag{2.13}$$

Or, resolution in the y-direction $\Delta R_y = \frac{D}{2}$ becomes

$$\Delta R_y = \frac{D}{2} = \frac{\lambda R\, \Delta f_i}{2V \sin\theta} \tag{2.14}$$

The above resolution is independent of antenna beamwidth (as was *not* the case in real array imaging radars) and depends primarily on attainable frequency resolution Δf_i of the signal processing equipment.

Note that for a matched transmit/receive system the duration of target illumination T should be $1/\Delta f_i$. This means that during the data acquisition time T the aircraft flies a length of $L = VT$ where V is its velocity. Thus, the equivalent antenna array length becomes

$$L = VT = \frac{V}{\Delta f_i} = \frac{\lambda R}{D \sin\theta} \tag{2.15}$$

Multiplying the numerator and denominator of (2.14) by T and using the value of L from (2.15) we get

$$\Delta R_y = \frac{D}{2} = \frac{\lambda R}{2L \sin\theta} \tag{2.16}$$

The value of D/2 given by the above equations represents the best attainable y-direction resolution and is used in calculations as a measure of merit. In actual practice, however, the y-direction resolution is obtained by a set of tuned filters as shown in Figure 2-4. This figure shows a set of filters with sin x/x drop-off characteristics.

The 3-dB bandwidths of the filters of Figure 2-4 are approximately one-half of the null-to-null bandwidth. Referring to Figure 2-4, a typical target will, most probably, fall into two

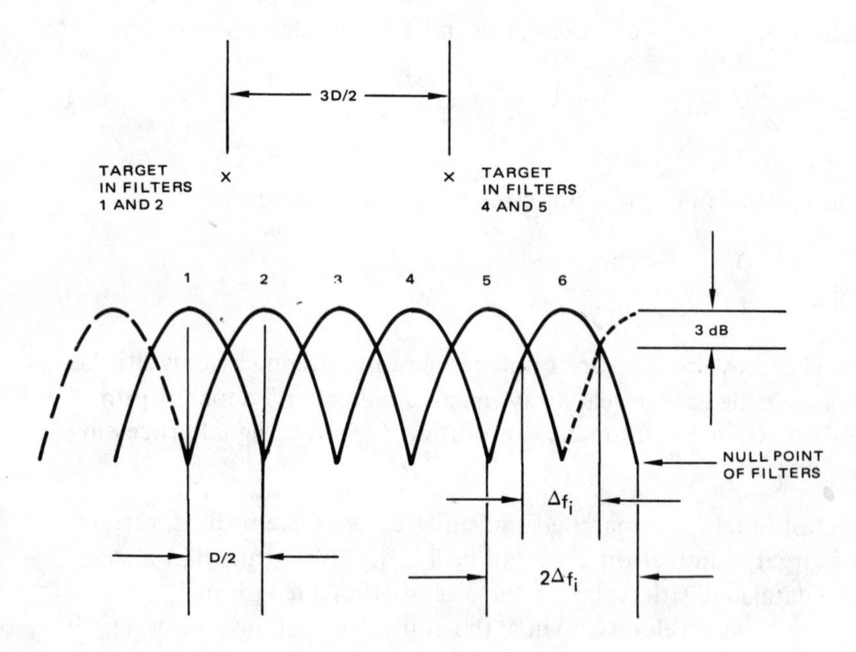

Figure 2-4 Actual and Theoretical Y-Direction Resolution in Synthetic Array Radars

adjacent filters, e.g., filters 1 and 2. Averaging will put the target in the center of these filters as shown. To obtain proper resolution separation filter number 3 should be skipped and the next resolvable target will probably appear in filters 4 and 5. Averaging will put the target between these filters as shown.

Thus, from Figure 2-4 the practical attainable resolution becomes 3D/2.

Figure 2-5 shows two ground maps obtained by two different airborne radars. The conventional map is obtained by an airborne radar equipped with a regular antenna dish pointed directly ahead and *not* a long antenna on the side of the aircraft for greater y-direction resolution. The doppler ground map is obtained with a radar equipped with the same size antenna as the conventional map but using synthetic array signal processing techniques. Note the remarkable improvement of the doppler ground map over the conventional ground map.

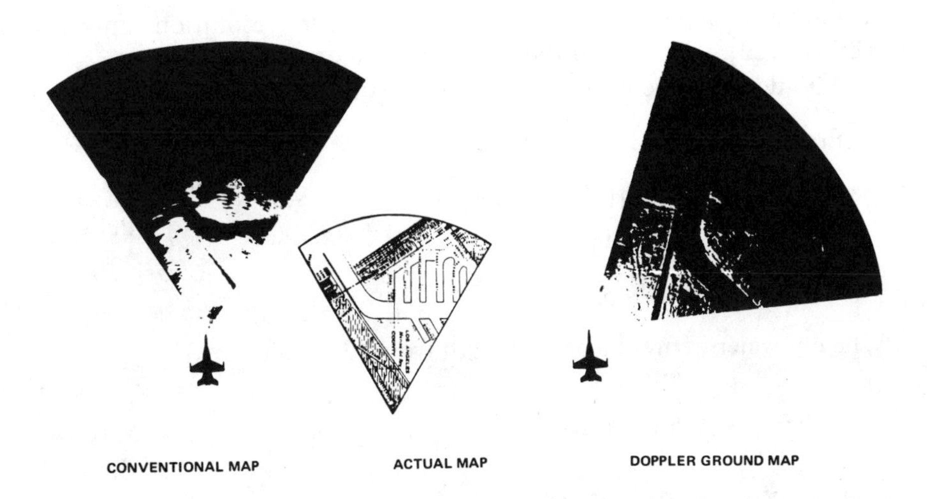

Figure 2-5 Conventional and Synthetic Array (Doppler) Ground Maps Obtained by an Airborne Radar

Another important observation that should be made about Figure 2-5 is the orientation of these ground maps. The conventional map represents the area directly ahead of the aircraft. The doppler ground map, on the other hand, represents the area to the side of the aircraft, as only resolution cells to the side of the aircraft can be separated by their doppler shifts. The resolution cells directly ahead produce negligible doppler shift separation because of small subtended angles between these resolution cells and the velocity vector of the aircraft and the fact that at these angles the cosine function remains practically constant very close to unity. As discussed before, the resolution of conventional radars is determined by their beamwidth and is independent of doppler shifts. For this reason areas in front of the radar can be mapped as long as the antenna illuminates such areas.

2.3.1 Example

Consider an aircraft equipped with a synthetic array radar flying with a velocity V = 150 meters per second mapping the ground at a range of 5 km. The signal processing resolution capability is $\Delta f_i = 20$ Hz and the wavelength is $\lambda = 1.5$ cm. For $\theta \cong 90$ degrees calculate y-direction resolution.

From equation (2.14) we have

$$\Delta R_y = \frac{D}{2} = \frac{(1.5\times10^{-2})(5\times10^{3})(20)}{2(150)} \tag{2.17}$$

$$= 5 \text{ meters}$$

The equivalent antenna array length from (2.15) becomes

$$L = \frac{150}{20} = 7.5 \text{ meters} \tag{2.18}$$

2.4 MAXIMUM RESOLUTION

Maximum resolution in the y-direction can be obtained by considering the fact that the antenna illumination coverage on the ground should be greater than the equivalent antenna array length as obtained above. This will insure the return of target energy from a point on the ground during the filter charge up time $T = 1/\Delta f_i$. Using the antenna beamwidth relationships from (1.9), we have

$$\frac{R\lambda}{\ell} > L = \frac{\lambda R}{D \sin\theta}, \tag{2.19}$$

where ℓ is the actual antenna length and L is the equivalent array length on the ground. Noting that $\Delta R_y = D/2$, equation (2.19) for $\theta \cong 90°$ results

$$\Delta R_y \geqslant \frac{\ell}{2} \tag{2.20}$$

That is, the best y-direction resolution obtainable is equal to one half of the actual antenna length ℓ.

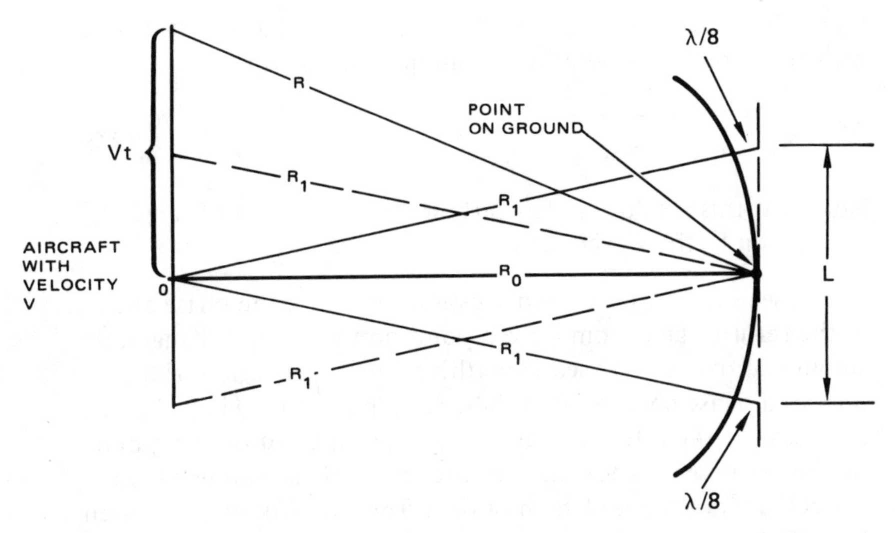

Figure 2-6 Range to a Point on the Ground as the Aircraft Flies with a Velocity V

2.5 FOCUSED AND UNFOCUSED ARRAYS*

As the aircraft flies the distance L, a point on the ground appears at distance R_1, R_o and R_1 again as shown in Figure 2-6. Since the phase of the arriving signal is proportional to the distance traveled, the phase of the arriving signal will change with time. Assuming a phase shift of 90 degrees or $\lambda/4$ is acceptable for flight distance of between zero to L, we can write the equation

$$(R_o)^2 + \left(\frac{L}{2}\right)^2 = (R_o + \lambda/8)^2$$

$$R_o^2 + L^2/4 = R_o^2 + \lambda R_o/4 + \lambda^2/64 \quad , \qquad (2.21)$$

where $\lambda/8$ is the acceptable *one-way* phase change. Neglecting the last term of equation (2.21) we can solve for L

$$L = \sqrt{\lambda R_o} \qquad (2.22)$$

*References 12 and 15.

Using the above expression in (2.16) and assuming $\sin\theta \cong 1.0$ we get the y-direction resolution for an unfocused array

$$\Delta R_y = \frac{D}{2} = \sqrt{\lambda R_o/2} \tag{2.23}$$

Note that this resolution depends on the square root of range R_o, in addition to the wavelength λ.

It is seen from the above discussion that since the phase angle of the return signal from a given point on the ground changes, the amount of data to be used from this point is a function of the tolerated phase change. As will be seen in the following discussion, the phase of the returned signal from a given point on the ground can be predicted. Knowing this, the return signal can be phase corrected and a longer length of data from the ground point used for signal processing. In synthetic array radars, this phase correction and adjustment is usually termed as focusing, and arrays using it are called *focused* arrays.

Again consider Figure 2-6, with the aircraft at distance Vt from zero and an aircraft to point on the ground range of R. The range R can be written in terms of R_o and Vt as follows

$$R = [R_o^2 + (Vt)^2]^{1/2} \tag{2.24}$$

Expansion of the above equation in series can be written as follows

$$R = R_o + \frac{1}{2R_o}(Vt)^2 - \frac{1}{8R_o^3}(Vt)^4 + \cdots \tag{2.25}$$

Neglecting all terms following the second term of this equation we get

$$R - R_o = \frac{1}{2}\frac{V^2t^2}{R_o} \tag{2.26}$$

Using equation (2.26) and noting that each wavelength λ is equivalent to a 2π phase shift we can write the equation for the phase shift ϕ as follows

$$\phi = 1/2\ (2\pi)\frac{V^2 t^2}{\lambda R_o} = \frac{\pi V^2 t^2}{\lambda R_o} \tag{2.27}$$

Note that the above value represents the *one-way* phase change. The *two-way* phase change will equal twice the value given in (2.27).

Equation (2.26) can also be derived from kinematics considerations of Figure 2-6. From this point of view, the ground point may be considered the origin, with vector R_o having a tangential velocity V. Tangential velocity V will result in a centripetal acceleration A_R along R_o of

$$A_R = \frac{V^2}{R_o} \tag{2.28}$$

Using this in equation for distance R in terms of initial distance and acceleration

$$R = R_o + \frac{1}{2} A_R t^2\ , \tag{2.29}$$

we get

$$R = R_o + \frac{1}{2}\frac{V^2}{R_o} t^2\ , \tag{2.30}$$

which can readily be put in the form of equation (2.26).

Figure 2-7 shows a plot of the above quadratic equation (2.27) together with phase adjustments, the first step being the subtraction of 2π radians, followed by the implementation of the periodic reference signal in every 2π slot. Finally, after thresholding, the white and black zone plate is obtained for optical processing. The reason for this will become clear after a discussion of holography and photographic zone plates in the following chapter.

Note that now all of the focused data can be used, while in the previously discussed unfocused case, only the portion of the data extending the $\frac{\pi}{2}$ radians could have been used.

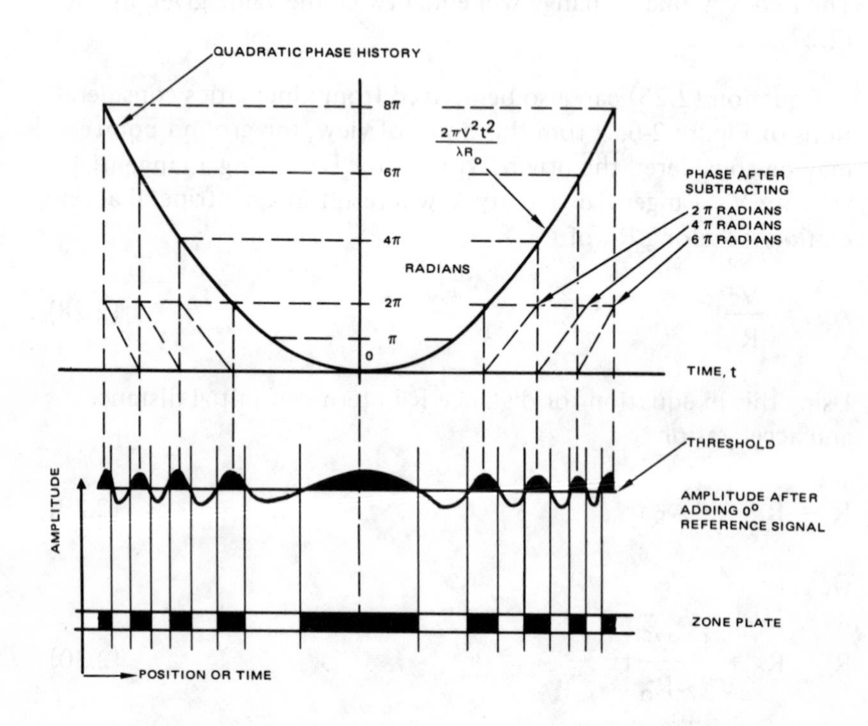

Figure 2-7 Quadratic Phase Angle from the Target with Appropriate Adjustments and Generation of Zone Plate

For the focused case the maximum achievable resolution, as discussed before, becomes $\Delta R_y = D/2 = \ell/2$ where ℓ is the physical dimension of the antenna.

Figure 2-8 shows a typical graph of y-direction resolution as a function of range on a logarithmic paper. The plot is for an antenna aperture of 5 feet and a wavelength of 0.1 feet. It is seen that the theoretical resolution of the focused case is not range dependent, while the resolution of the unfocused case is range dependent, i.e., square root of range. The conventional case resolution

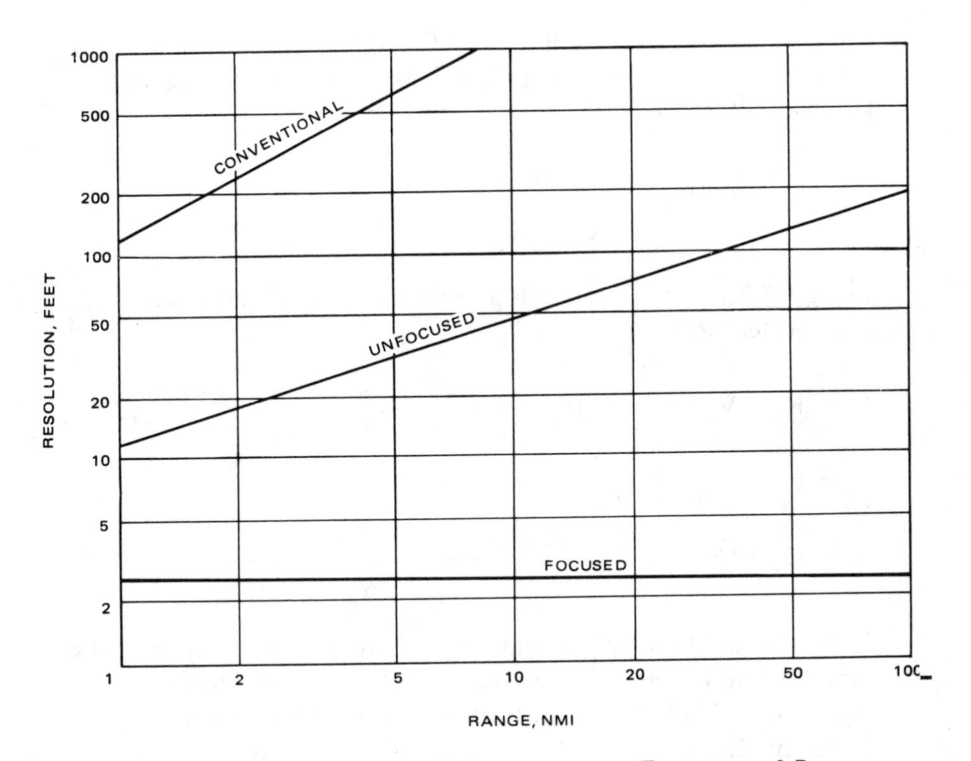

Figure 2-8 Resolutions in the Y-Direction as a Function of Range for the Cases Focused, Unfocused and Conventional Radars

is a function of beamwidth coverage on the ground and is directly proportional to range. Note that Figure 2-8 gives only the y-direction resolution versus range relationship. In a typical mapping radar system, range enters into several additional considerations such as power requirements, ground mapping region coverage, signal to noise ratio, and the quantity of signal processing.

2.6 SUMMARY AND EXTENSION OF EQUATIONS

With the above background we can easily derive relevant equations for the case where the antenna look angle is not perpendicular to the flight path or, equivalently, the mapping or squint angle is not 90 degrees. Figure 2-9 gives the flight geometry for a squint angle of θ. The components of velocity V along the vector R_0

and perpendicular to it will be $V\cos\theta$ and $V\sin\theta$, respectively. The centripital acceleration along R_o will be obtained by using tangential velocity $V\sin\theta$ as follows

$$A_R = \frac{(V\sin\theta)^2}{R_o} \tag{2.31}$$

Using the relation for range as a function of initial range, velocity and acceleration

$$R = R_o + V_R t + \frac{1}{2} A_R t^2 , \tag{2.32}$$

we get

$$R = R_o + V\cos\theta \; t + \frac{1}{2} \frac{(V\sin\theta)^2}{R_o} t^2 \tag{2.33}$$

Using the fact that the 2π phase shift represents the equivalent distance of one wavelength λ, we can transform the distances of equation (2.33) to phase shifts by multiplying both sides of (2.33) by $2\pi/\lambda$

$$\phi(t) = \phi(t_o) + \frac{2\pi}{\lambda} V\cos\theta t + \frac{\pi}{\lambda} \frac{(V\sin\theta)^2}{R_o} t^2 , \tag{2.34}$$

where $\phi(t)$ and $\phi(t_o)$ are substituted for $2\pi R/\lambda$ and $2\pi R_o/\lambda$, respectively. For $\theta_o = 90°$, equation (2.34) reduces to (2.27) where $\phi(t) - \phi(t_o) = \phi$ represents the phase shift from R_o to R. The derivative with respect to time of the second term on the right of equation (2.34), i.e., the term involving t, represents the projection of velocity along R_o or, equivalently, the doppler shift along R_o.

The doppler shift term is removed in both focused and unfocused synthetic array radar processing. The quadratic term involving t^2 is removed in focused signal processing methods resulting in better resolution capability.

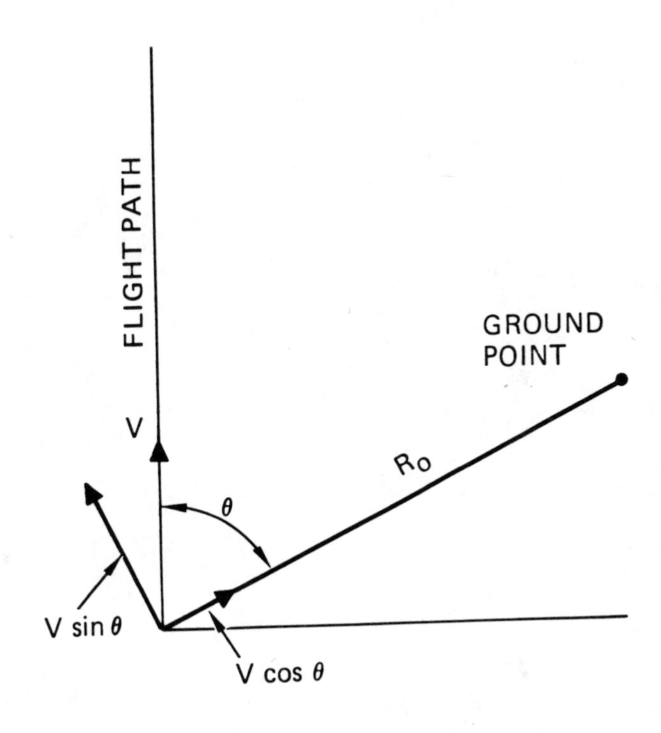

Figure 2-9 Flight Geometry for a Squint Angle of θ

The data of Figure 2-7 forms the basis of optical data processing in mapping radars. The ideas used in these processing methods are borrowed from holography, photographic zone plates and so on. In the next chapter, before we discuss optical processing methods of radar data we will give a brief discussion of holography and applicable optical procedures.

3 Signal Processing Methods

The signal processor of the synthetic array radar must store the incoming coherent returns for each range resolution element across the map width, preserving amplitude and phase of the returns at the highest doppler frequency of interest. After a synthetic array length of data has been stored, the processor must perform the filtering operation discussed previously.

Two general classes of synthetic array processors are currently in operation. They are the *optical* and the *electronic* processors. The focused optical processor remains the best processor for wide swath, high resolution applications where immediate access to map is not required. The electronic processors, both digital and analog, provide a greater range of operating modes and the possibility of immediate access to the mapped imagery.

Before discussing synthetic array processors, we will give a basic description of holography and photographic zone plates. These principles and methods are utilized in optical processors.

3.1 HOLOGRAPHY AND PHOTOGRAPHIC ZONE PLATES

Holography was originally conceived in 1947 by Hungarian-British scientist, Nobel Laureate Dennis Gabor as a means of producing three-dimensional images through the use of photographic plates and coherent lights*. Figure 3-1 illustrates the recording of data for future hologram production**. The scene is illuminated by a coherent laser light with resulting reflections towards the

---

*Coherence means that phase and frequency stability is achieved.

**Hologram is derived from Greek words holo (complete) and gram (message).

photographic plate. In addition, a reference laser light also directly illuminates the plate. Combination of the reflected light from the object and the reference light form a complicated wave pattern which is recorded on the photographic plate.

Now, the developed photographic plate can be used for the reconstruction of the original photographed objects by exposing it to the reconstructing laser light as shown in Figure 3-2. To a viewer the original objects appear in their three-dimensional forms.

Gratings and zone plates are also produced through the use of interfering signals. Figure 3-3 shows a set of photographic zone plates produced by sum and difference of interfering signals. Additions result in dark spots while cancellations result in light spots. The bottom figure shows the two-dimensional diagram of the zone plate.*

Another set of photographic zone plates can be constructed from the interference of plane and spherical waves as shown in Figure 3-4a. The bottom line of the figure has a focusing distance $f = n\lambda$, where n is an integer and λ is the wavelength with each line increasing in length by $\lambda/2$, half a wavelength. Note the unequal distances of light and dark zones. By using this zone plate and exposing it to a horizontal reconstructing wave set, three sets of waves will result. One set travels straight through horizontally; another acts as though it were diverging from the source point of the original spherical wave; and the third is a set which converges towards a point on the opposite side of the recorded circular pattern. We will see in the following discussion that the zone plates of Figure 3-4 are similar to synthetic array recordings of ground points.**

---

*References 11 and 13.

**These zone plates are also called Fresnel zone plates.

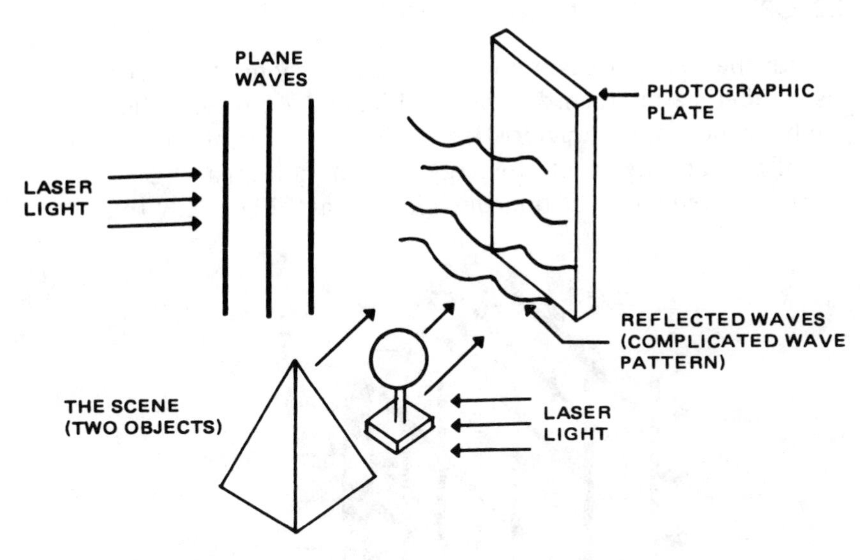

Figure 3-1 Hologram Construction on a Photographic Plate

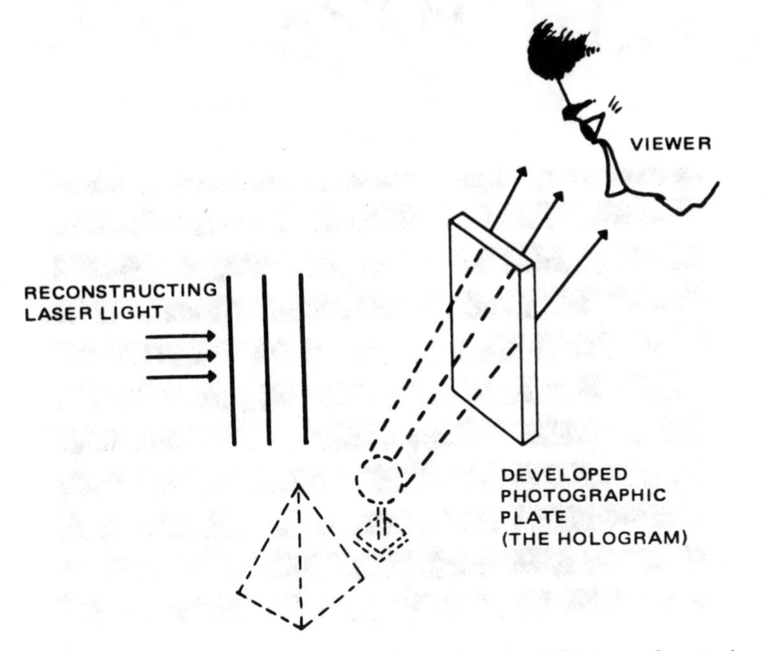

Figure 3-2 Reconstruction of the Original Objects from the Hologram

3.2 OPTICAL SIGNAL PROCESSING

With the above discussion it becomes an easy matter to see that the synthetic array recorded data of Figure 2-7 possesses the photographic zone plate characteristics of Figure 3-4a. Note that both sets of data are the result of energy emanating from a single source point. The reduction of the zone plate data of Figure 2-7 to a

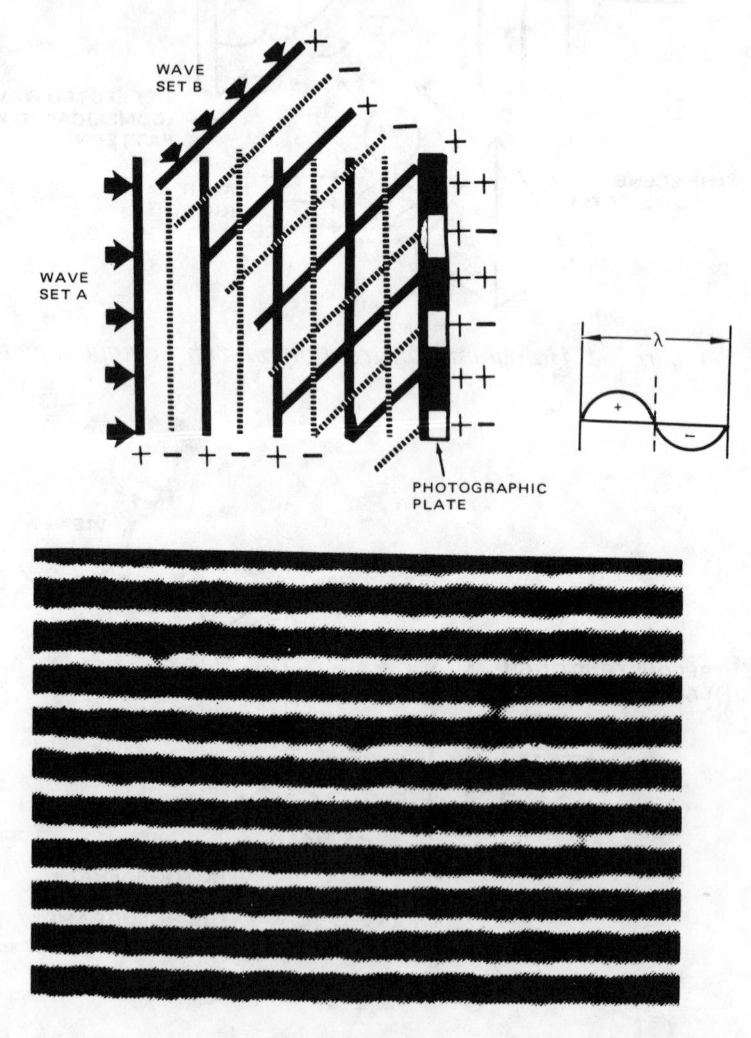

Figure 3-3 Production of Photographic Zone Plates by Interfering Signals

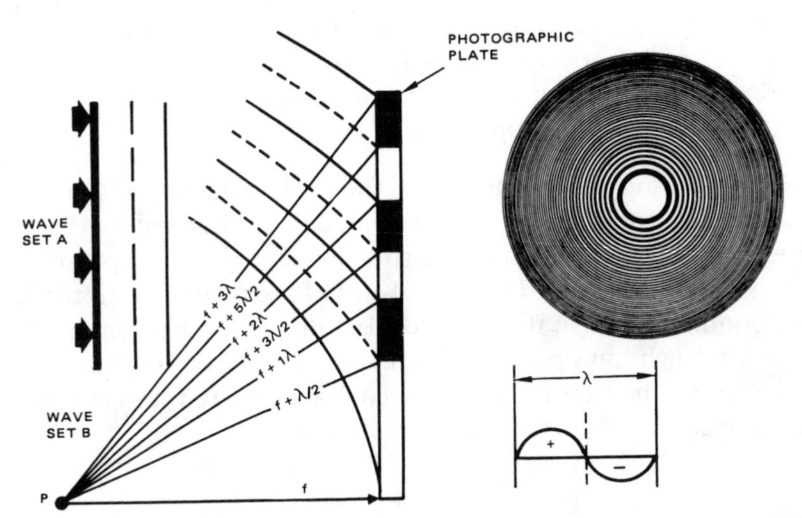

a) CONSTRUCTION OF A PHOTOGRAPHIC ZONE PLATE USING PLANE AND SPHERICAL WAVES. HEAD-ON VIEW IS SHOWN WITH REDUCED SCALE.

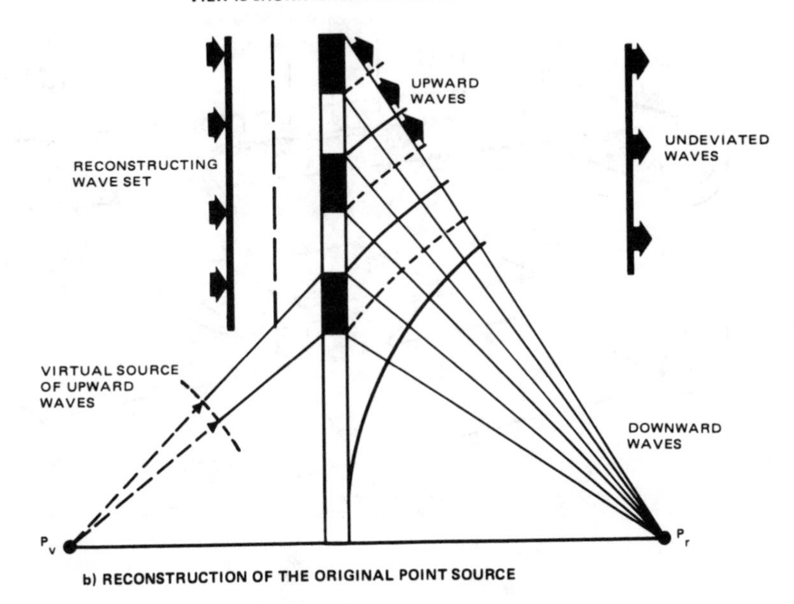

b) RECONSTRUCTION OF THE ORIGINAL POINT SOURCE

Figure 3-4 Construction/Reconstruction of Photographic Zone Plates Using Plane and Spherical Waves

single point is implemented by a coherent reference reconstructing reference light as was done in the case of Figure 3-4b. This procedure for a synthetic array radar is shown in Figure 3-5. This figure shows the reconstruction of a real image from zone plate data obtained from a ground point.

Figure 3-6 shows flight geometry and radar recording of synthetic array data for two near and far ground points covered by the antenna beam. The same figure also shows an optical computer for processing of this data. The conical lens of the figure is used to take out differences in the recorded data due to radar target range, making the light rays parallel. The balance of the lens system shown brings the light produced by conical lens into focus in the output plane as two distinct ground points.

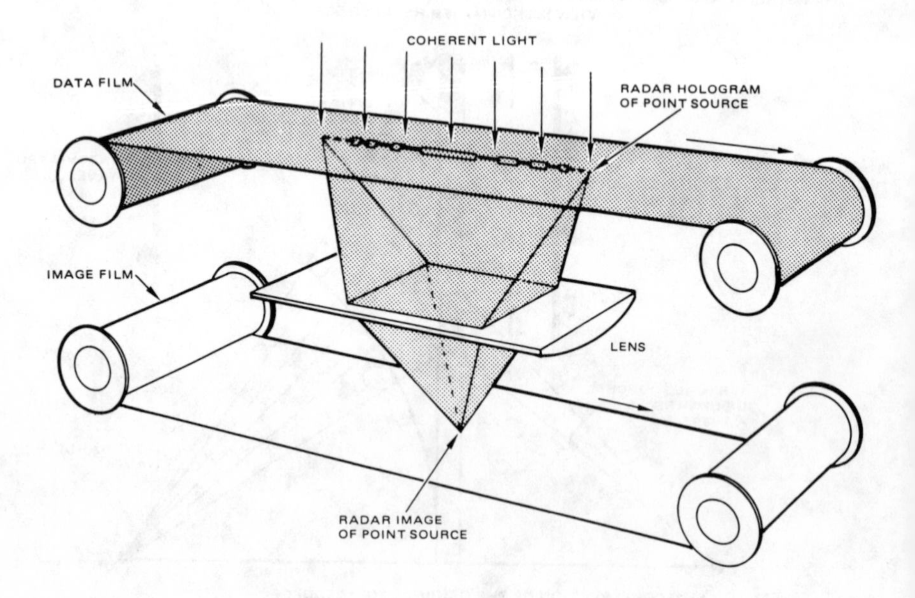

Figure 3-5 Holographic Image Produced from Recorded Zone Plate Data

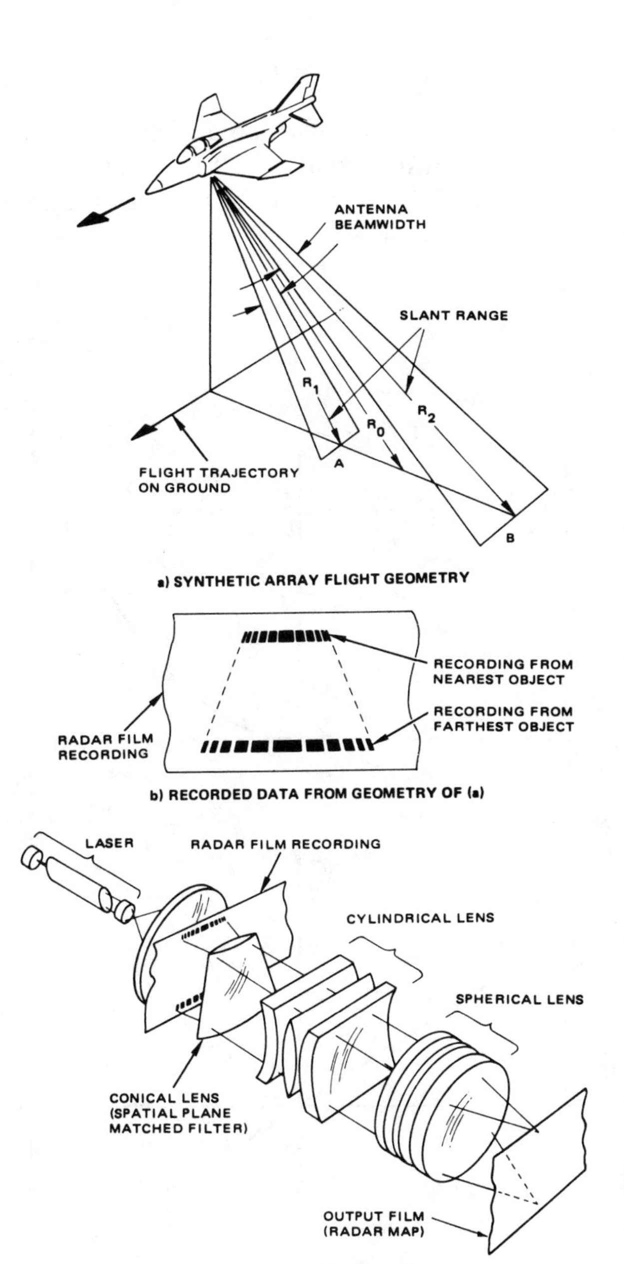

Figure 3-6 Optical Data Processing from Synthetic Array Recorded Data

3.3 ELECTRONIC SIGNAL PROCESSING

Consider the quadratic phase diagram of Figure 2-7 repeated in Figure 3-7 for the following discussion. The quadratic phase equation as given by (2.27) is written for two way transmission/reception as

$$\phi = \frac{2\pi V^2 t^2}{\lambda R_o} \tag{3.1}$$

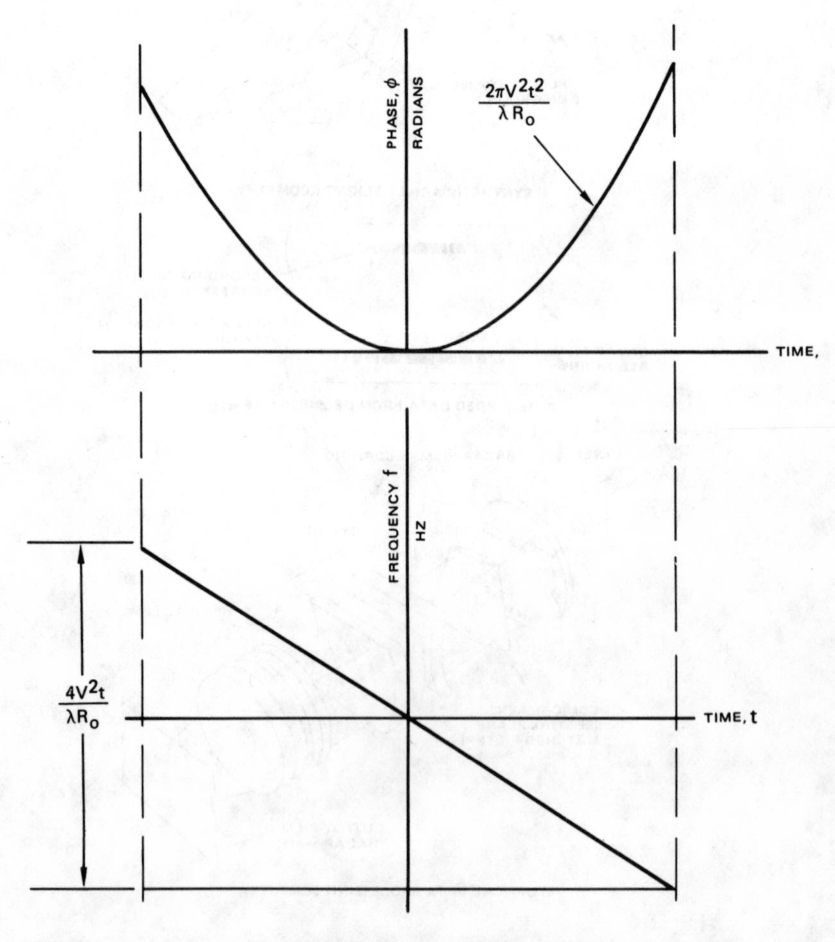

Figure 3-7 Phase and Frequency Excursions of a Single Point on the Ground as a Function of Time

The rate of change of phase with respect to time $d\phi/dt$ is angular frequency ω. Using this, from (3.1) we have

$$\omega = 2\pi f = d\phi/dt = \frac{4\pi V^2 t}{\lambda R_o} \qquad (3.2)$$

Solving these equations for f we get

$$f = \frac{2V^2 t}{\lambda R_o} \qquad (3.3)$$

The two sided frequency excursion will be twice this value as shown in Figure 3-7.

Having Figure 3-7 in mind let us consider Figure 3-8. In this figure an aircraft flies with a velocity V and has a beamwidth which can illuminate six points on the ground. In position A, the radar begins to illuminate point 1 and at position B it completes the illumination of this point. At position A the component of aircraft velocity along the vector A-1 will give rise to the frequency shift in the frequency-time diagram. As the aircraft flies from A to B this frequency will decrease passing through zero as shown in the bottom figure. Other points of the figure will also have the same frequency history with the total frequency excursion equalling $4V^2 t/\lambda R_o$.

If focusing of the returned signal is *not* used, because of overlapping frequency lines, only the portion of the slanted frequency excursions of Figure 3-8 bounded by T can be used for each point -3, -2, -1, . . . as input data. Using the value $t = T/2$ in the two-sided frequency excursion equation

$$\Delta f = \frac{4V^2 t}{\lambda R_o} \qquad (3.4)$$

will result in a

$$\Delta f = \frac{2V^2 T}{\lambda R_o} \qquad (3.5)$$

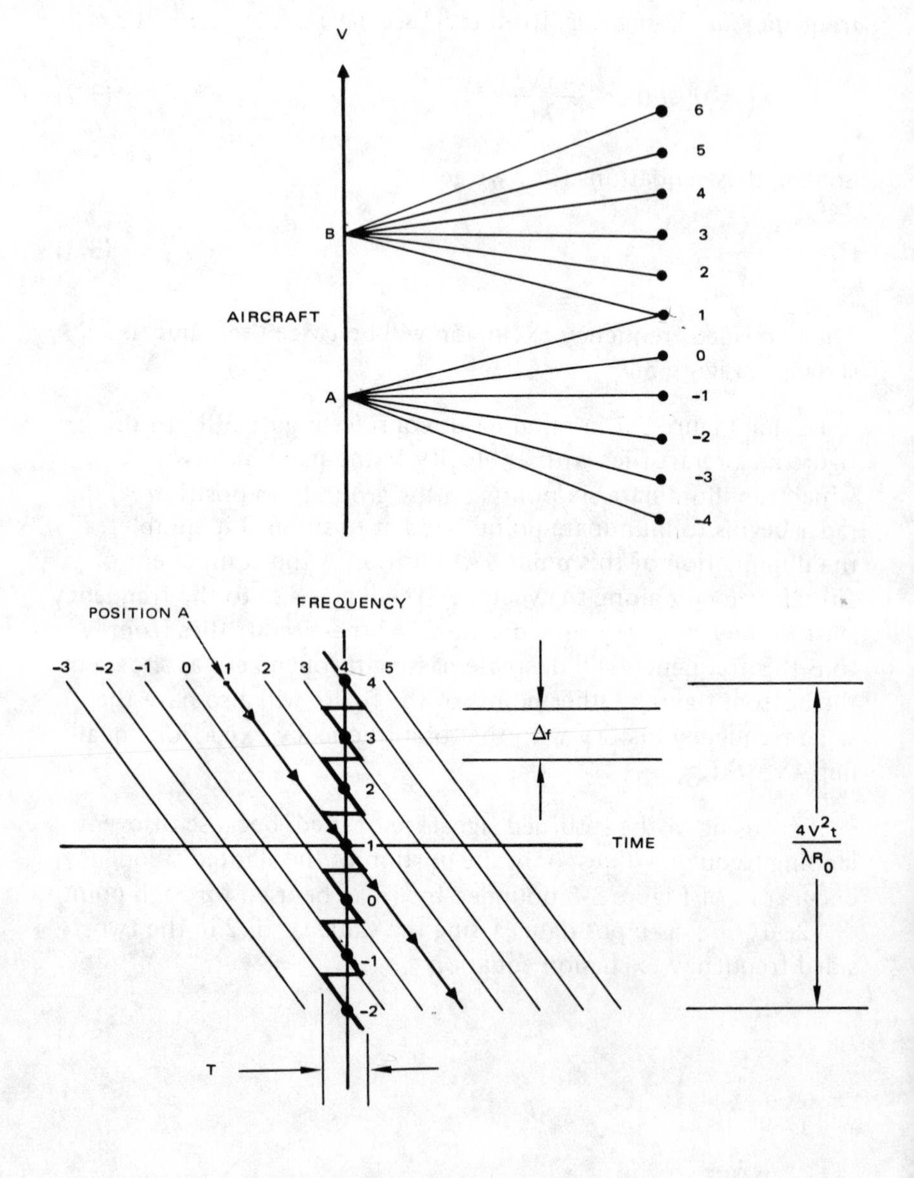

Figure 3-8 Aircraft Flight and Frequency History of Points on the Ground

This frequency bandwidth is equivalent to a time of T seconds ($\Delta f = 1/T$). Using this in (3.5) and utilizing $L = VT$ we get

$$L = VT = 0.707\sqrt{\lambda R_o}\quad, \tag{3.6}$$

where L is the flight path. Using (3.6) in (2.16) of the previous chapter, we get the unfocused y-direction resolution as

$$\Delta R_y = \frac{D}{2} = 0.707\sqrt{\lambda R_o} \tag{3.7}$$

Note that the unfocused resolution was derived previously by phase shift considerations and given in equation (2.23) as

$$\Delta R_y = \frac{D}{2} = 0.50\sqrt{\lambda R_o} \tag{3.8}$$

The difference between the coefficients 0.707 and 0.50 of equations (3.7) and (3.8) comes from assumptions made in deriving these equations. Inherent in the derivation of (3.7) is the assumption that the bandwidth $\Delta f = 1/T$, which does not necessarily correspond to the 90 degree phase shift assumed in deriving equation (3.8). In obtaining (3.6) had we used $\Delta f = 2/T$, the null-to-null bandwidth, we indeed would have come out with $L = \sqrt{\lambda R_o}$ and $\Delta R_y = \sqrt{\lambda R_o}/2$ as given in equations (2.22) and (2.23) of Chapter II. Equation (2.23) is given in this chapter as (3.8).

From Figure 3-8 reducing Δf ($=1/T$) or equivalently increasing y-direction resolution, will require focusing of the returned signal much the same way as given in Figure 2-7 of the previous chapter. In the case of Figure 3-8 focusing will effectively allow the utilization of frequency excursions from various ground points without interference between frequency returns of these points.

Focusing can be effected in one of two ways. In one method, matched filters similar to the pulse compression filters of Chapter V can be used. Each filter has time/frequency characteristics such as to match one of the slant frequency lines starting at points -3, -2, -1, . . . of Figure 3-8. Each filter, in effect, follows one of these lines and puts out a burst of power at the appropriate terminating frequency.

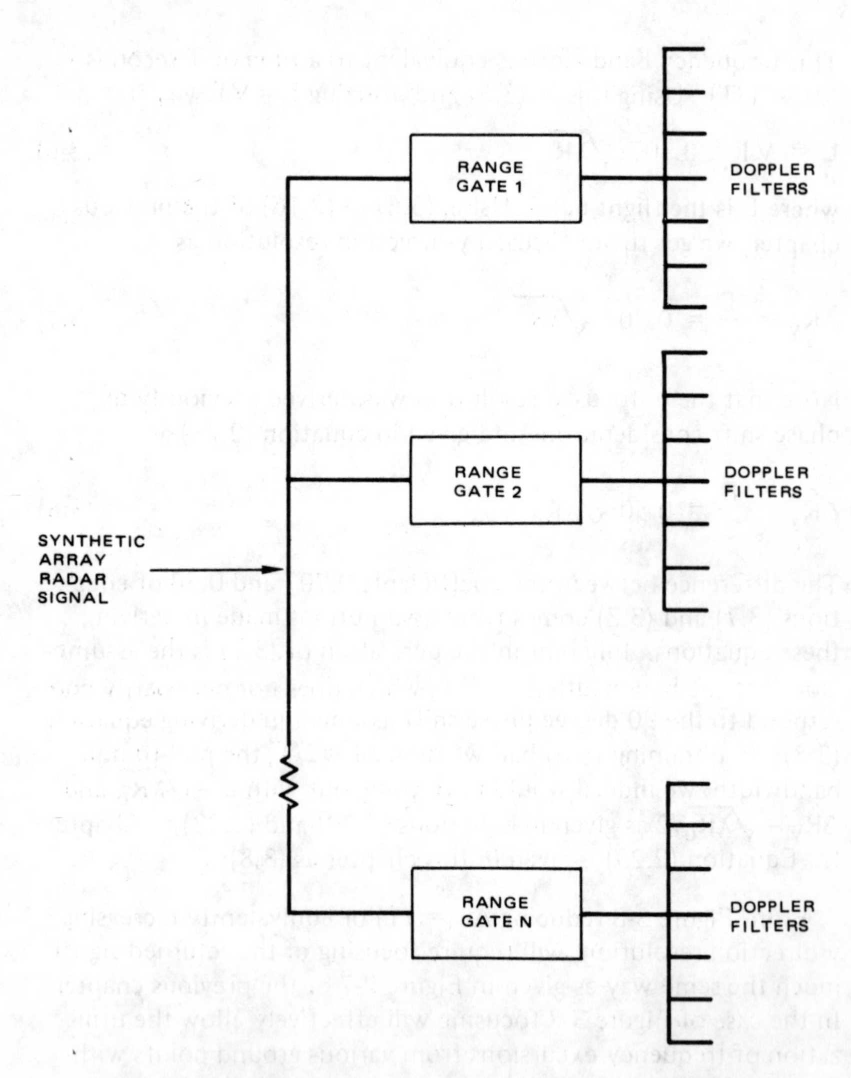

Figure 3-9 Range Gates Followed by Doppler Filters for Electronic Data Processing

In the second method, a slanted time/frequency reference signal is used which, in effect, rotates the slanted frequency excursions of Figure 3-8 such that they become parallel to the time axis of the figure. Each frequency line then intercepts the frequency axis at designated frequencies -2, -1, 0, 1, 2, ... This operation is implemented through a swept oscillator with the same slope as the target return signal. The target return signal of points -3, -2, -1, . . . is mixed with the swept oscillator signal producing a constant frequency signal for each target return.

With the above background it becomes a simple matter to see that the electronic signal processing of synthetic array radars can be accomplished by a set of range gates, for x-direction resolution, each range gate followed by a set of doppler filters for the y-direction resolution as shown in Figure 3-9.

Note that instead of analog filtering, represented by doppler filters of Figure 3-9, the spectral signal processing can also be accomplished by Fourier transform and digital computers.

3.4 ELECTRONIC SIGNAL PROCESSING CONSIDERATIONS*

Figure 3-10 shows the 3-dB beamwidth coverage and associated maximum and minimum ranges to the ground in the vertical plane.

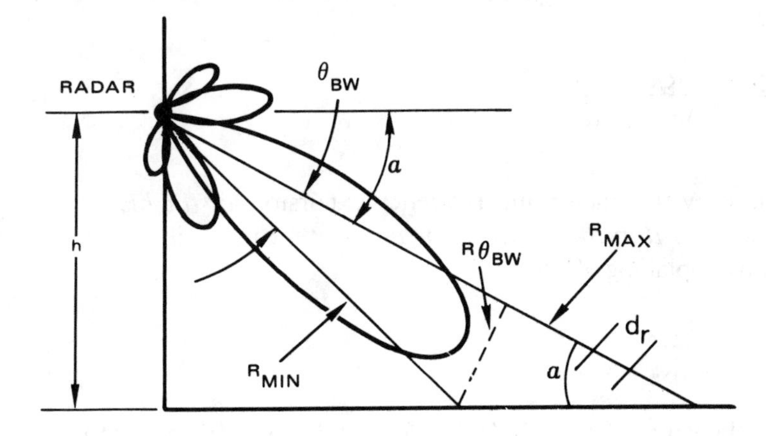

Figure 3-10 Maximum and Minimum Range-to-Ground for the 3dB Antenna Beamwidth θ_{BW}

*Reference 5.

The range resolution element d_r is also shown on the figure. The number of range gates or range bins can be calculated as follows

$$N_R = \frac{R_{max} - R_{min}}{d_r}$$

$$= \frac{R\lambda}{\ell\, d_r \tan\alpha}, \quad (3.9)$$

where N_R is the number of range gates, and $d_r = c\tau/2$ is the range resolution along R with τ being the pulse duration and c velocity of propagation. The second equation gives the value of N_R in terms of an average range R value and the antenna beamwidth $\theta_{BW} = \lambda/\ell$, where ℓ is the antenna dimension and λ is the wavelength.

Equation (3.9) can also be written in terms of resolution along x-direction ΔR_x by using the relation $d_r = \Delta R_x \cos\alpha$. This will result in

$$N_R = \frac{R\lambda}{\ell\, \Delta R_x \sin\alpha} \quad (3.9a)$$

Similarly the maximum frequency excursion, *or the highest frequency within the synthetic array length,* can be obtained from (3.5) by replacing VT = L

$$(\Delta f)_L = \frac{2VL}{\lambda R_o} \quad (3.10)$$

Using the value of $L = \lambda R/D$ (for θ = 90 degrees) from (2.15) we get

$$(\Delta f)_L = \frac{2V}{D} \quad (3.10a)$$

From Nyquist's sampling theorem the number of time samples for representing a given frequency should be at least twice that frequency, i.e., the number of samples per second in this case will be $2(\Delta f)_L$. Since these samples are taken within the time that it takes to fly a synthetic array length L; the total number of samples will be

$$N_D = 2(\Delta f)_L T$$

$$= \frac{4V}{D} \cdot \frac{\lambda R}{VD}$$

$$= \frac{4\lambda R}{D^2}, \qquad (3.11)$$

where the value of $T = 1/\Delta f_i$ is obtained from (2.13) with $\theta = 90$ degrees.

The total number of samples for all range gates N_R will be

$$N = \text{total number of samples}$$

$$= N_R N_D$$

$$= \frac{(R_{max} - R_{min})}{d_r} \frac{4\lambda R}{D^2}, \qquad (3.12)$$

where N_R is obtained from (3.9). Using the value of N_R from (3.9a), and denoting the azimuth resolution D/2 by ΔAZ, we get another form of equation (3.12) as follows

$$N = \frac{R\lambda}{\ell (\Delta R_x) \sin \alpha} \cdot \frac{R\lambda}{(\Delta AZ)^2} \qquad (3.12a)$$

From this equation it is seen that the number of samples is directly proportional to the square of distance R, and inversely proportional to resolution in range and azimuth directions.

Equations (3.12) and (3.12a) are used in determining the signal processing requirements of synthetic array radars. This will be discussed in Chapter V.

3.4.1 PRF Selection

An important parameter in the design of synthetic array radars is the bounds of the pulse repetition frequency (PRF) of the transmitted signal. The limits of the PRF are determined by the mapping range of the radar and the doppler frequency shift of the extreme points of the antenna coverage. From Figure 3-11 the doppler shift between points A and B is given by

$$f_d = \frac{2V}{\lambda} \cos\theta - \frac{2V}{\lambda} \cos\left(\theta + \theta_{BW}\right), \tag{3.13}$$

where θ_{BW} is the 3-dB antenna beamwidth. Using small angle approximations for θ_{BW} the above equation reduces to

$$f_d = \frac{2V}{\lambda} \theta_{BW} \ (\sin\theta) \tag{3.14}$$

For $\theta \cong 90$ degrees, and using the relation $\theta_{BW} = \lambda/\ell$ where λ is the wavelength and ℓ is the antenna length, we get

$$f_d = 2V/\ell \tag{3.15}$$

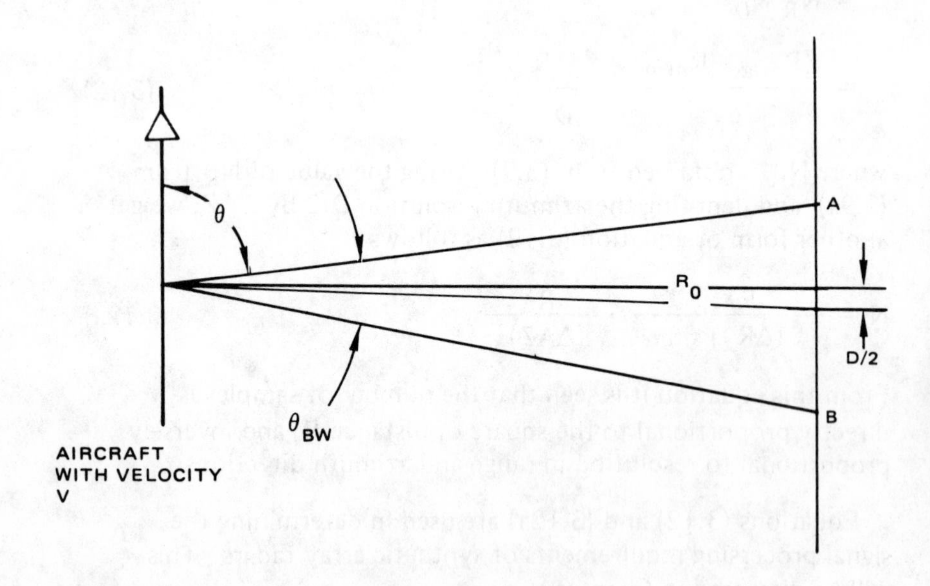

Figure 3-11 Ground Coverage in the Y-Direction for a 3dB Beamwidth θ_{BW}

The frequency representation of a train of pulses with interpulse period T is given in Figure 3-12. Assuming that the doppler shift excursion of equation (3.15) should fit around center frequency f_o of the figure, we get

$$f_d \leqslant PRF \quad (3.16)$$

$$PRF \geqslant 2V/\ell \quad (3.17)$$

This equation insures that the ground return from other PRF lines will not interfere with the return from transmitted frequency f_o.

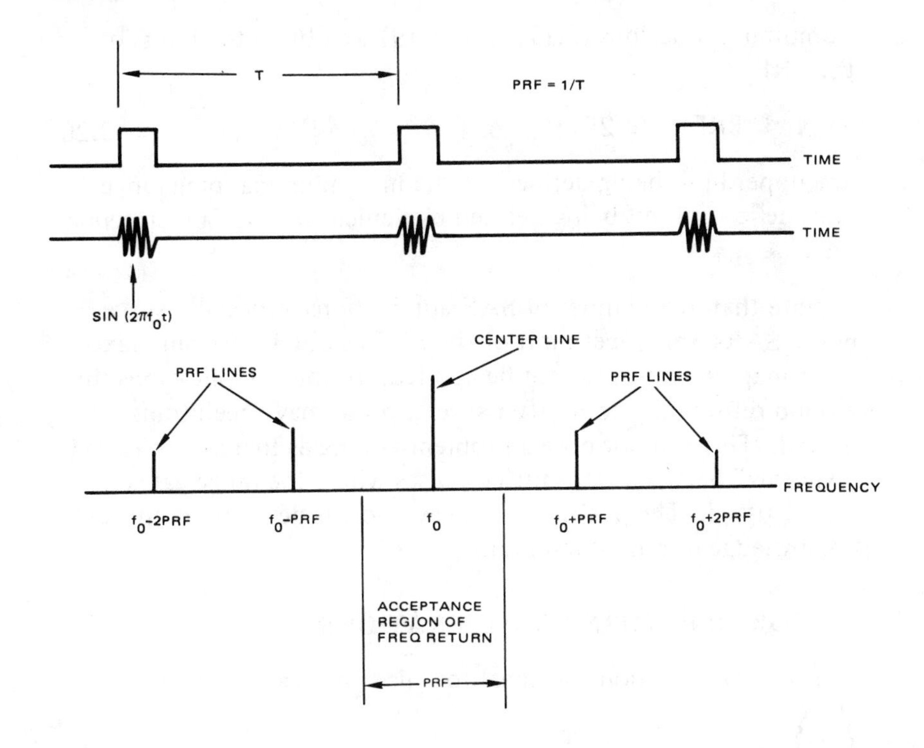

Figure 3-12 Transmitted Pulses and Spectral Diagrams of PRF Lines

The upper limit of the PRF is attained from a consideration of the maximum mapping range and the fact that the returned pulse from this range should come within the interpulse period T. This condition will insure that there will not be a range ambiguity condition. In equation form this becomes

$$T \geqslant \frac{2R_{max}}{c} , \tag{3.18}$$

where R_{max} is the maximum mapping range and c is the velocity of propagation. Using the relation $T = 1/PRF$ in (3.18) we get

$$PRF \leqslant c/2R_{max} \tag{3.19}$$

Combining equations (3.17) and (3.19) we obtain the limits of the PRF

$$2V/\ell \leqslant PRF \leqslant c/2R_{max} , \tag{3.20}$$

the upper limit being defined by the maximum mapping range and the lower limit being defined by vehicle velocity and antenna size.

Note that in a number of SAR applications, especially spaceborne SARs, the upper limit of the PRF as obtained from maximum mapping range cannot be applied. In these applications the ground return may come after several pulses have been transmitted. This will not pose a problem as long as there is a clear "window" between transmitted pulses where the range gates can be fitted. The position of this window is determined through a knowledge of radar-target range.

3.5 POWER RETURN CONSIDERATIONS

The radar equation for signal to noise ratio can be written

$$\left(\frac{S}{N}\right)_p = \frac{PG^2 \lambda^2 \sigma}{(4\pi)^3 R^4 (kTW) L} , \tag{3.21}$$

where

$\left(\frac{S}{N}\right)_p$ = signal to noise ratio per pulse

P = peak pulse power

G = antenna gain

R = range to target

k = Boltzmann's constant

T = equivalent receiver noise temperature

N = receiver noise

= kTW

W = receiver noise bandwidth

L = system losses

λ = transmit wavelength

σ = radar target cross section

The radar target cross section of the resolvable area on the ground is given by using equations (1.4) and (2.14) as follows

$$\sigma = \left[\left(\frac{D}{2}\right)\left(\frac{c\,\tau}{2\cos\alpha}\right)\right]\sigma_o \quad , \tag{3.22}$$

where σ_o is the backscattering coefficient and depends on the type of the terrain illuminated by the radar. In addition, σ_o depends on the angle of incident α. Typical backscattering coefficients for ocean and concrete roadways using an antenna with vertical polarization are given in Figure 3-13. It is noted that the coefficients, in addition to depending on the angle of incident α, also depend on transmitted frequency.*

At incident angles of 20-40 degrees the backscattering coefficient of land areas with large perturbations such as forests is considerably greater than that of smooth surfaces such as concrete roadways and various sea states. Figure 3-13b shows about a 10 dB difference between backscattering coefficients of smooth (concrete) and rough (gravel) roadway surfaces. The difference

*Reference 5, chapter 6.

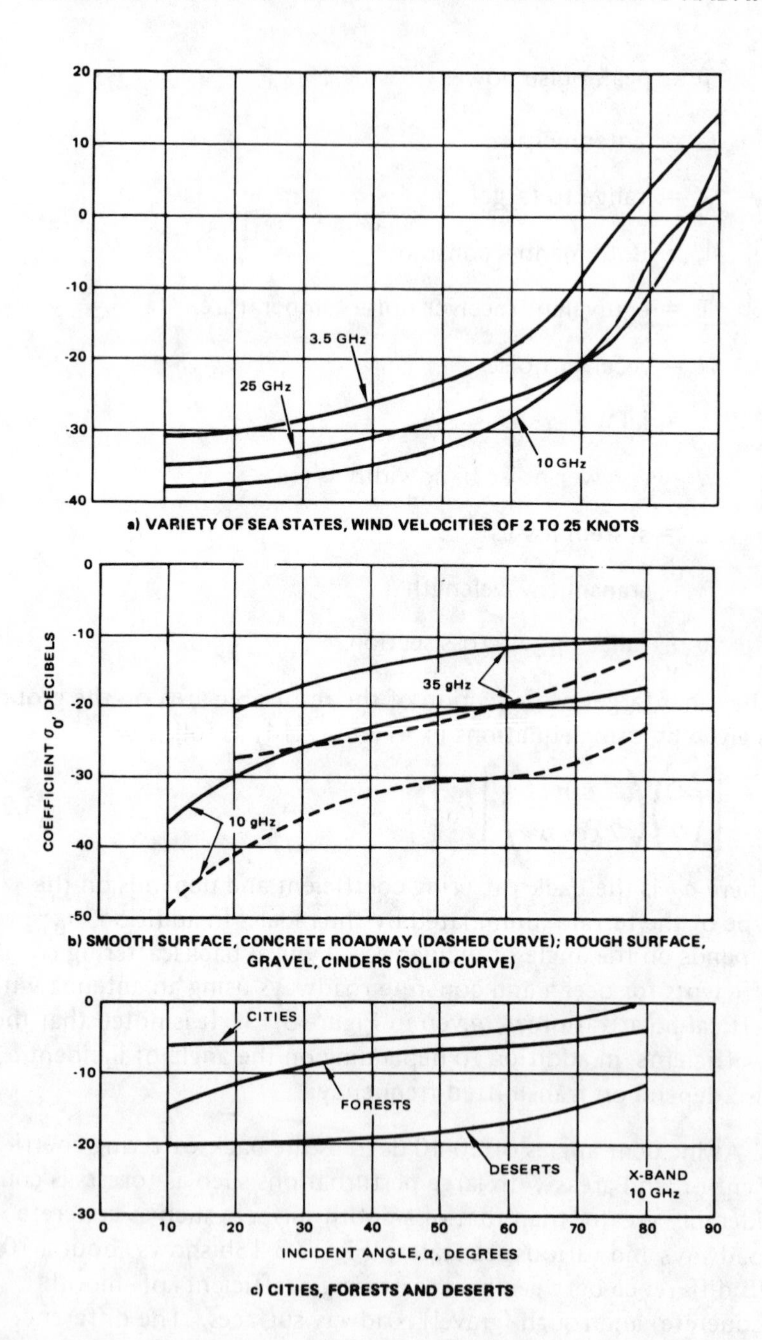

Figure 3-13 Backscattering Coefficients for Oceans and Other Surfaces

between backscattering coefficients of concrete roadways and forests at x-band, from Figures 3-13b and c, is about 25dB, i.e., a factor of over 300 in power return. Buildings, bridges and other structures which present surface cut-offs and discontinuities to the incident radar energy present large target radar cross sections and thereby high power returns.

From the above discussion it can be seen that factors affecting radar return power include: surface roughness relative to the wavelength of the illuminating signal, angle of incident of radar energy, geometry and orientation of the object with respect to the incident radar wave, and the polarization of transmitting and receiving antennas. The difference in the return energy of a given point on the ground when received from two different illuminating positions in space is sometimes used in SAR radars to enhance the quality of the radar imagery. By using this *stereo* process, SAR radars can extract additional features of the illuminated area on the ground, such as geometric height measurements.

As an example of calculating the value of target cross section σ from (3.22) consider a radar operating at 60 degree look-down angle with x- and y-direction resolution of 15 meters each. For an x-band transmission the value of σ_o for sea state can be obtained from Figure 3-13a. This value for 60 degree and $\lambda = 3.0$ cm (10 GHz) becomes

$$10 \log \sigma_o = -28; \quad \sigma_o = 1.58 \times 10^{-3} \tag{3.23}$$

Using this in (3.22) we get the radar target cross section as

$$\sigma = (15)(15)(1.58 \times 10^{-3}) = 0.356 \text{ square meter} \tag{3.24}$$

The above target cross section for a forest background from Figure 3-13c becomes

$$10 \log \sigma_o = -6; \quad \sigma_o = 0.251 \tag{3.25}$$

and

$$\sigma = (15)(15)(0.251) = 56.47 \text{ square meters} \tag{3.26}$$

This example shows a 150 fold difference in target cross section due to mapping terrain below.

The above calculation in conjunction with Figure 3-13 can be used for calculating background target return power levels. The presence of protruding objects within this background will, of course, cause additional power returns. For example, cars traveling on desert-crossing highways and ships at sea will represent large radar target cross sections to incident electromagnetic energy. Table 3-1 gives very approximate radar cross sections for a number of moving and non-moving structures which may be used in power return calculations.

TABLE 3-1 RADAR TARGET CROSS SECTIONS

OBJECT	CROSS SECTION M^2
Jeeps, Cars	50
Trucks, Small Freighters	500
Bridges, Medium Freighters	1000
Large Buildings, Warehouses, Hangars	10000

Figure 3-14 gives a set of backscattering coefficients for a low grazing angle of 5.5 degrees, vertically polarized antenna and a transmission frequency of 16.4 GHz. This figure was obtained from a paper by C. R. Griffin and published in the proceedings of the Synthetic Aperture Radar Technology Conference March 8-10, 1978, Las Cruces, New Mexico. There seems to be general agreement between values given in Figure 3-14 and those of previously discussed Figure 3-13 for the few comparable conditions of these figures.

3.5.1 Atmospheric Absorption*

The earth's atmosphere causes propagation loss of electromagnetic energy due to molecular absorption by oxygen and water

*James Martin, *Communication Satellite Systems*, Prentice Hall, NJ, 1978, Pages 110-115.

vapors present in the atmosphere. In addition, weather conditions such as fog and clouds, rain and snow also cause absorption of radar energy.

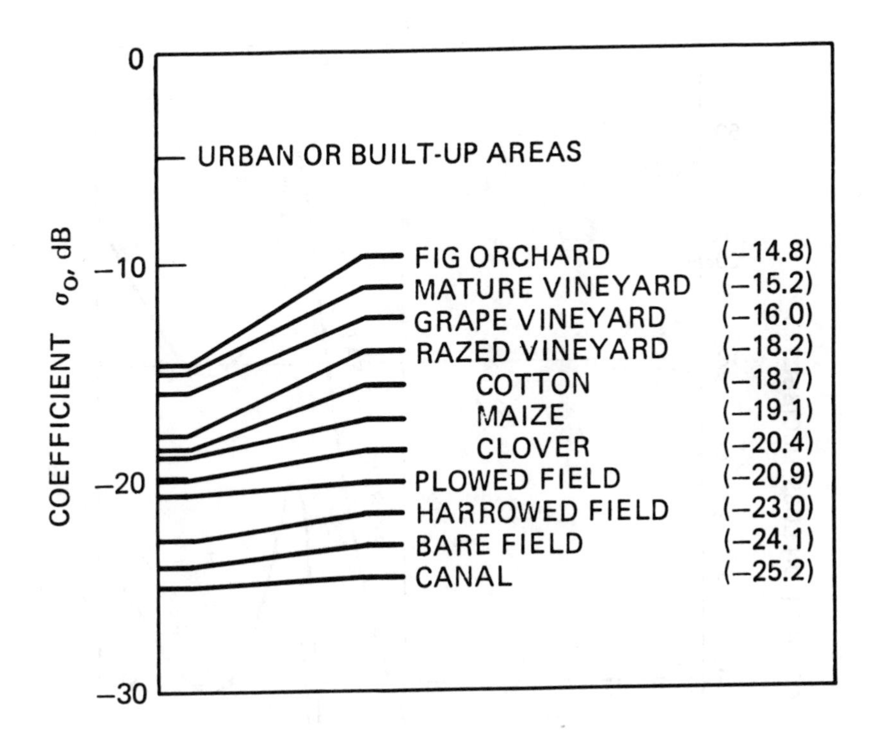

Figure 3-14 Backscattering Coefficient at Low Angle of Incidence (5.5 Degrees) and Frequency of 16.4 GHz

Basically, absorption of electromagnetic energy in the atmosphere is caused by the collision of electrons contained in the radio waves with those in the earth's atmosphere. This collision results in the loss of electromagnetic energy that is absorbed by the atmosphere.

Figures 3-15 and 3-16 show the maximum absorption of electromagnetic energy, due to oxygen and water vapor, in traveling through the atmosphere. These figures are drawn for ground based radars using elevation angle as a parameter. In the case of spaceborne radars, the values given in these figures can be used to approximate the actual atmospheric loss. Note that the absorption is greater for small angles of elevation because the radio waves have a longer path to travel through the atmosphere.

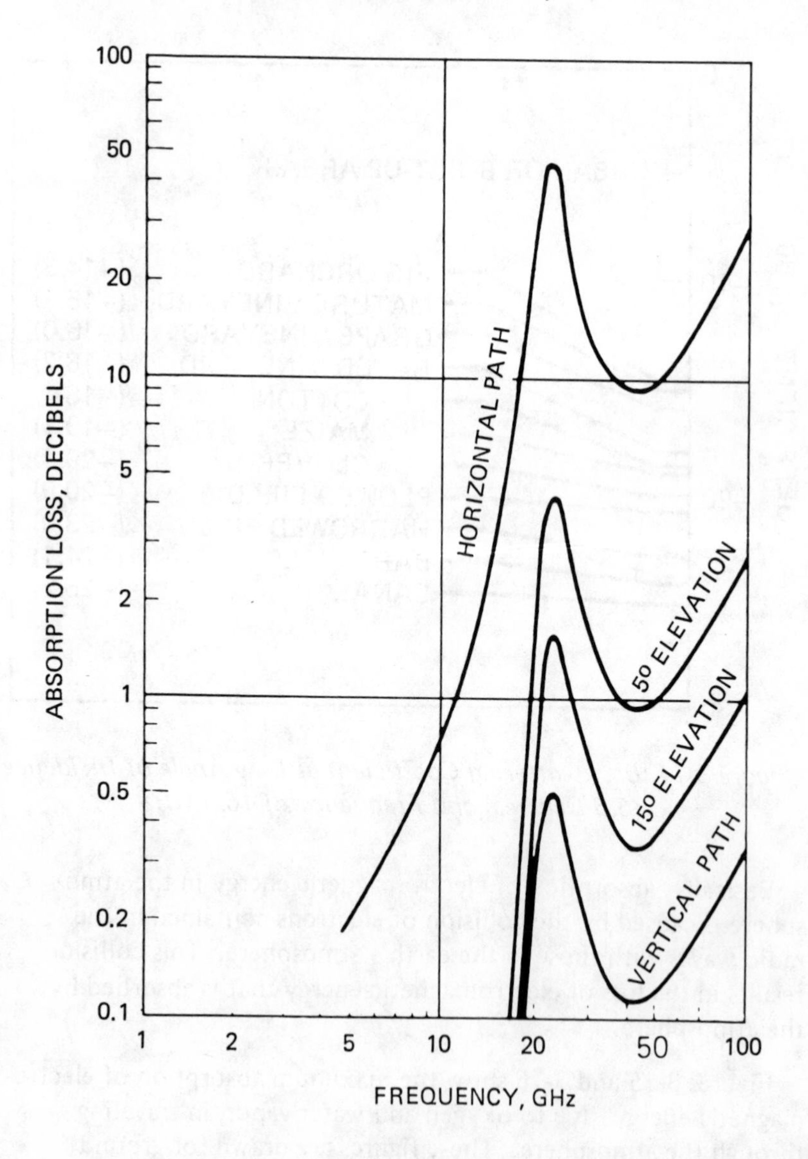

Figure 3-15 Absorption in the Atmosphere Caused by Water Vapor (Radar at Sea Level)

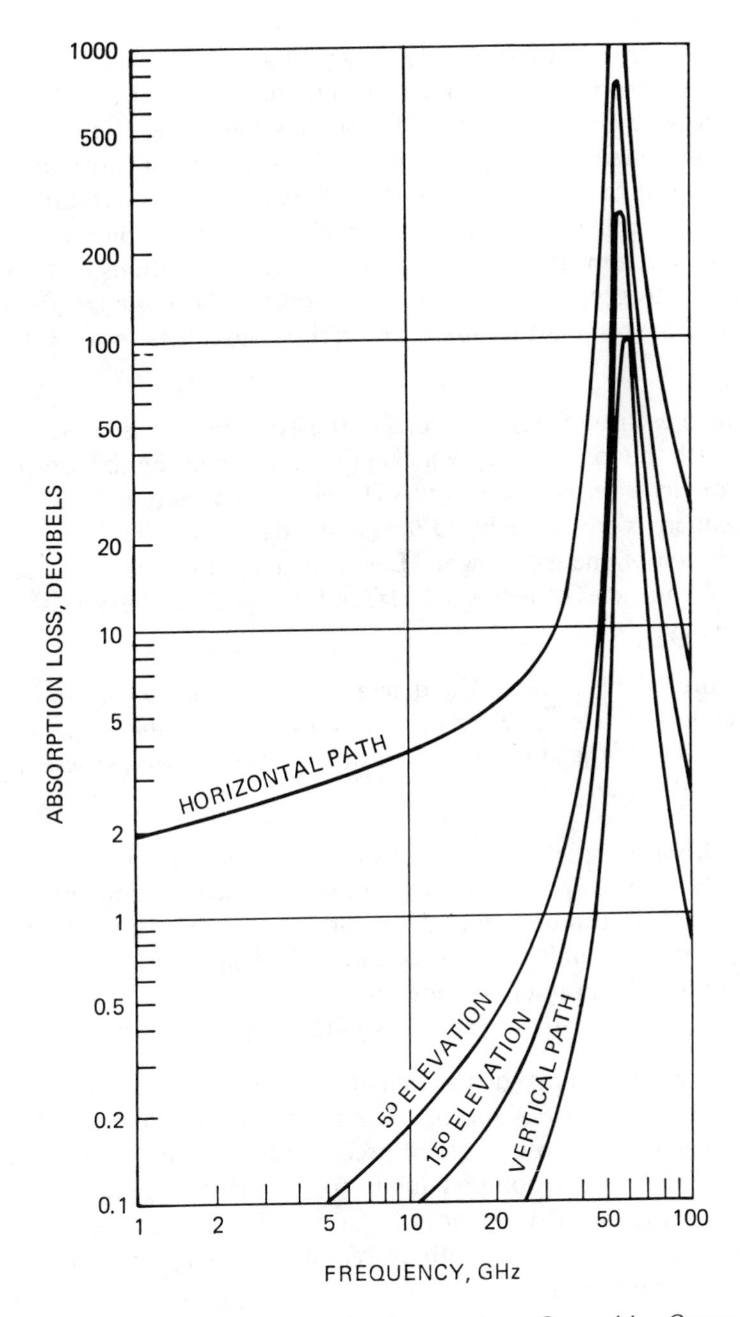

Figure 3-16 Absorption in the Atmosphere Caused by Oxygen (Radar at Sea Level)

The absorption by molecular oxygen has a sharp peak at 60 GHz and absorption by water molecules has a peak at 21 GHz. This peak absorption is caused by radio waves changing the rotational energy levels of oxygen molecules, causing molecular resonants at these frequencies. This resonance and the resulting build-up of molecular vibrational amplitude feed on the transmitted electromagnetic energy, causing high absorptions. Atmospheric nitrogen does not have a resonant peak at these frequencies and the resonant frequency of carbon dioxide is above 300 GHz.

The presence of free electrons in the atmosphere (as is the case with the ionosphere) will also result in atmospheric absorption of electromagnetic energy. The electron density of the ionosphere is greatly reduced during the dark hours, reducing absorption of energy at night. Electron absorption mainly affects radio frequencies of below 1 GHz, and is negligible above this frequency.

Figure 3-17 is a composite diagram showing the absorption caused by electrons and oxygen and water vapor combined. At about 60 GHz the atmospheric absorption is high, while there is a dip at about 30 GHz.

In airborne radars, where electromagnetic energy travels through a limited portion of the atmosphere, attenuation curves of Figure 3-18 can be used. These curves show the atmospheric loss per nautical mile of distance traveled. Thus, the total atmospheric loss can be obtained by multiplying the values given in this figure by the two-way radar-target range.

Weather conditions such as rain and fog also adversely affect radar power returns. Atmospheric attenuation due to rain will, of course, depend on the rate of precipitation. Figure 3-19 gives atmospheric attenuation loss due to fog per nautical mile of traveled distance. Note in both Figures 3-19 and 3-20 the attenuation loss increases with decreasing wavelength (increased transmit frequency).

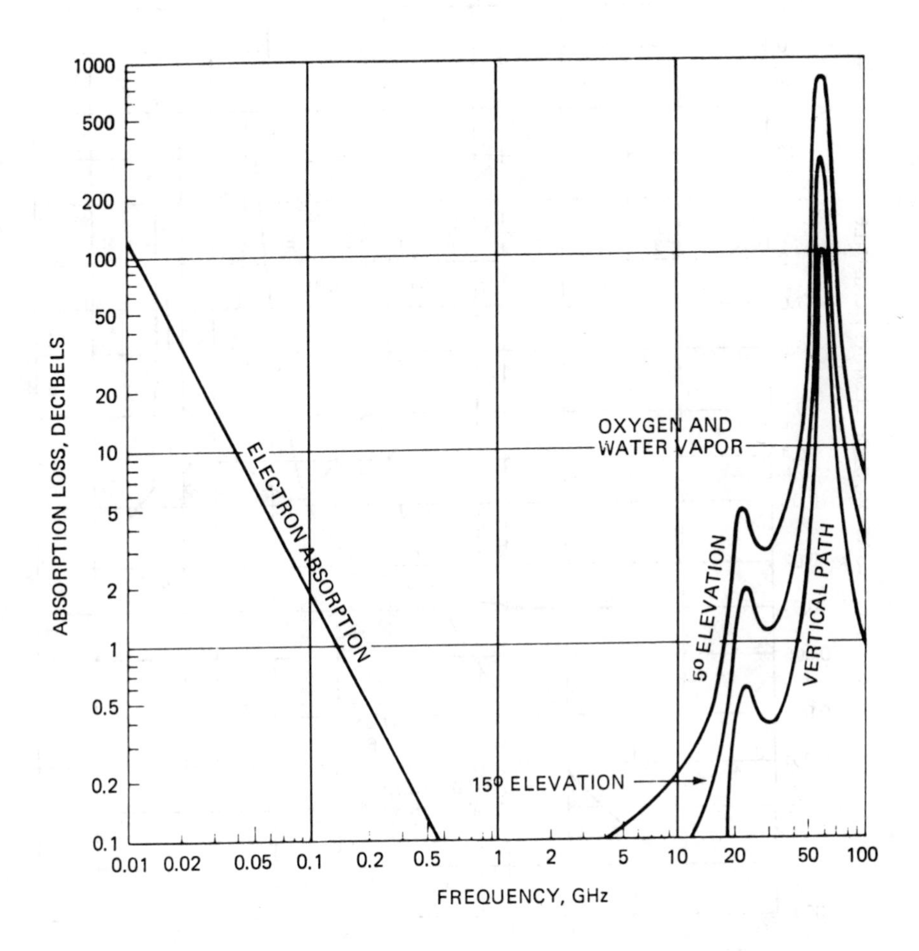

Figure 3-17 Absorption in Atmosphere by Free Electron and Combined Oxygen and Water Vapor (Radar at Sea Level)

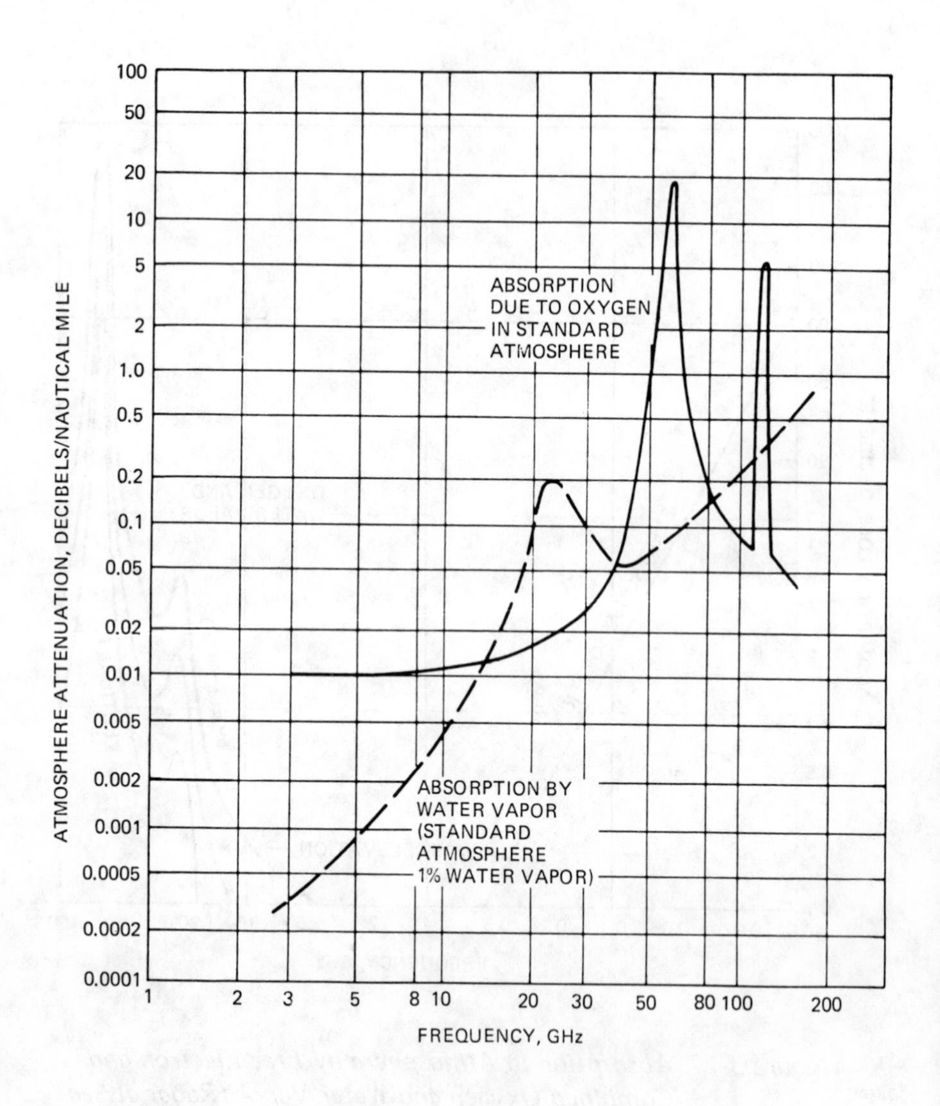

Figure 3-18 Attenuation of Electromagnetic Energy by Atmospheric Gases at Standard Atmosphere

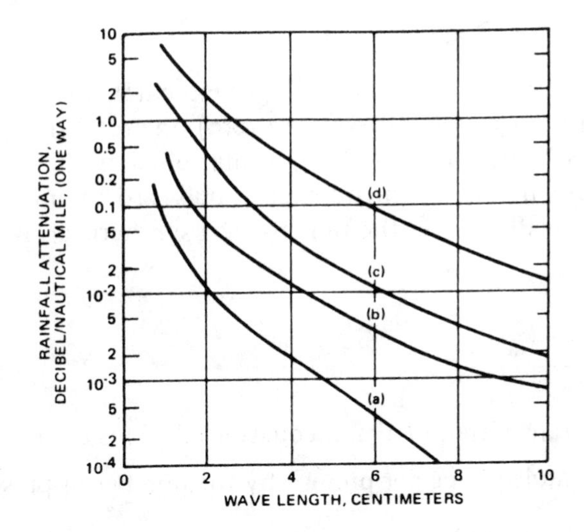

Figure 3-19 Rain Attenuation as a Function of Wave Length a) Drizzle – 1/4 mm/hr, b) Light Rain – 1 mm/hr, c) Moderate Rain – 4 mm/hr, d) Heavy Rain – 16 mm/hr

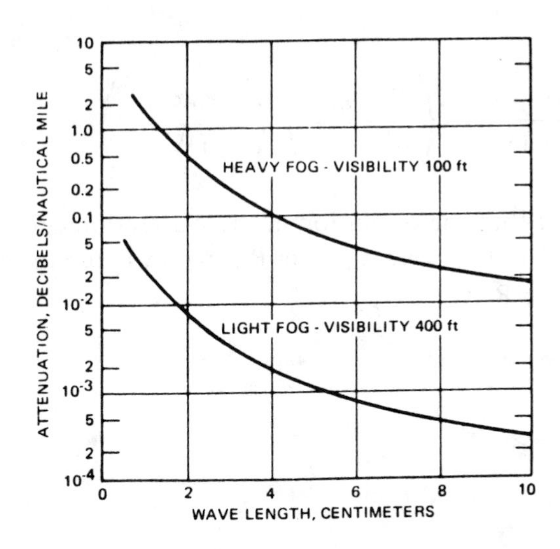

Figure 3-20 Attenuation of Electromagnetic Energy Due to Fog – Light Fog 0.032 g/m^3, Heavy Fog 2.3 g/m^3

3.5.2 Average Power Requirements

If the signals returned from the small ground patch are coherently added (integration loss equal to zero), the total signal-to-noise ratio (S/N) of equation (3.21) is multiplied by the number of returned pulses. The number of pulses to be integrated is the radar PRF times the time to fly a synthetic array

$$f_r T = f_r L/V = \frac{f_r R \lambda \csc\theta}{DV}, \tag{3.27}$$

where f_r is the transmitted PRF and equation (2.15) is used for the value of L. Multiplying per pulse $\frac{S}{N}$ by the number of pulses of equation (3.27) we get

$$\frac{S}{N} = \left(\frac{S}{N}\right)_p f_r T \tag{3.28}$$

Using equations (3.21), (3.22) and (3.27) in (3.28) we get the total $\frac{S}{N}$;*

$$\frac{S}{N} = \frac{PG^2\lambda^2}{(4\pi)^3 R^4 (kTW)L}\left(\frac{Dc\tau\sigma_o}{4\cos\alpha}\right)\left(\frac{f_r R\lambda\csc\theta}{DV}\right) \tag{3.29}$$

Utilizing the fact that the resolution in the x-direction $\Delta R_x = c\tau/2\cos\alpha$ and also, the average power $\overline{P}$ can be written in terms of the peak power P as

$$\overline{P} = \frac{P\tau}{T'} = \frac{Pf_r}{W}, \tag{3.30}$$

where $\tau = 1/W$ is the pulse width and $T' = 1/f_r$ is the interpulse period, we can write equation (3.29) in terms of average power $\overline{P}$

*Note that in equation (3.27) L represents the radar losses.

$$\overline{P} = \frac{(4\pi)^3 R^3 kTL}{G^2 \lambda^2 \sigma_0 \Delta R_x} \left(\frac{2V}{\lambda}\right)\left(\frac{S}{N}\right) \sin\theta \tag{3.31}$$

The above result gives the average power requirements of a synthetic array radar. The following should be observed:

1) Power is proportional to R^3.

2) Power is proportional to increased x-direction resolution ΔR_x.

3) Power is proportional to velocity V.

4) Power is independent of y-direction resolution D/2.

From equation (3.31), increased resolution in the x coordinate, i.e., small ΔR_x values, will require increased transmitted power. In a number of synthetic array radars this is achieved by pulse compression techniques. In this method a long pulse, or more average power, is transmitted and upon return the pulse is compressed to obtain the desired ΔR_x resolution. The pulse compression technique will be discussed in detail in Chapter V.

3.5.3 Power Return with Pulse Compression*

As mentioned before, the pulse compression technique of improving x-direction resolution is discussed in Chapter V of this book. The power return calculations using pulse compression are somewhat different than the ones for an uncompressed pulse, as discussed above. Since we have been discussing power return and S/N calculations, for continuity, we will derive the power return equations using pulse compression in this section. The full understanding of the pulse compression method, however, may not come until the discussion in Chapter V.

Starting from per pulse signal to noise ratio equation (3.21) we have

$$\left(\frac{S}{N}\right)_p' = \frac{PG^2\lambda^2\sigma}{(4\pi)^3 R^4 (kTW)L} \tag{3.32}$$

*The author wishes to acknowledge the assistance of Mr. Melvin Whipple in the derivation of equations of this section.

Assuming an initial uncompressed pulse duration of τ_i and a compressed pulse duration of τ_c we can write the pulse compression ratio D as

$$D = \frac{\tau_i}{\tau_c}, \tag{3.33}$$

The signal voltage improvement factor (Figure 5-4 of Chapter V) will be $\sqrt{D}$, and the signal power improvement factor will be D.

From (3.27) the total number of pulses integrated over coherent integration time will be

$$N_p = \frac{f_r R \lambda}{2\delta_{az} V \sin \sigma}, \tag{3.34}$$

where δ_{az} represents azimuth or y-direction resolution. Using the value of radar target cross section in terms of azimuth and range resolution cells δ_{az} and δ_r, where δ_r is resolution along the x-direction, we get

$$\sigma = \delta_{az}\, \delta_r\, \sigma_o, \tag{3.35}$$

where σ_o is the backscattering coefficient. Finally, using the relation for the average power $\overline{P}$ as a function of peak power P of equation (3.32) we have

$$\overline{P} = P\tau_i f_r \tag{3.36}$$

or

$$P = \frac{\overline{P}}{\tau_i f_r} \tag{3.37}$$

Using relations (3.33) through (3.37) in (3.32) we get

$$\frac{S}{N} = \left(\frac{\overline{P}}{\tau_i f}\right)(G^2\lambda^2)(\delta_{az}\delta_r\sigma_o)\left(\frac{\tau_i}{\tau_c}\right)\left(\frac{f_r R\lambda}{2\delta_{az} V}\right) \Big/ (4\pi)^3 R^4 (kTW) L \sin\theta, \tag{3.38}$$

where $\frac{S}{N}$ represents the total signal to noise ratio.

Simplification of (3.38) will result

$$\frac{S}{N} = \frac{\overline{P}G^2\lambda^3\delta_r\sigma_o}{(2V)\tau_c(4\pi)^3} \times \frac{1}{(kTW)LR^3\sin\theta} \tag{3.39}$$

Note that $W = \frac{1}{\tau_c}$ where W is the receiver bandwidth and τ_c is duration of the compressed pulse. Using this, the quantity $\tau_c W$ becomes equal to one. Thus, equation (3.39) becomes

$$\frac{S}{N} = \frac{PG^2\lambda^3\delta_r\sigma_o}{(2V)(4\pi)^3 kTLR^3\sin\theta} \tag{3.40}$$

The above relation can be readily put in the form of equation (3.31)

$$\overline{P} = \left(\frac{(4\pi)^3 R^3 kTL}{G^2\lambda^2\sigma_o\Delta R_x}\right)\left(\frac{2V}{\lambda}\right)\left(\frac{S}{N}\right) \sin\theta\ , \tag{3.41}$$

where the value of δ_r is substituted by ΔR_x. Note that equations (3.31) and (3.41) do not differ. Thus, the pulse compression method does not reduce average power requirements of the radar.

4 Implementation and Application of SAR

To obtain high quality imagery, careful attention must be paid to sidelobes that are introduced in the storage process and the filter formation process. In addition, sidelobes are introduced by such factors as uncompensated non-linear aircraft motion and phase variations of the transmitted and received signals through the antenna, transmitter, receiver, and atmosphere. In the initial formulation of a synthetic array radar design, an error budget must be drawn up to describe the amount of acceptable final error and its allocation among the various contributors. Generally accepted criteria for high quality imagery are peak sidelobes to be at least 25 dB below the mainlobe and integrated sidelobes to be at least 15 dB below. For the purposes of comparison, the first or the highest sidelobe of a standard sin x/x filter pattern is about 13 dB (power) below the mainlobe, and integrated sidelobes of this pattern are about 9 dB below the mainlobe.

The effects of integrated and peak sidelobes are quite different. Peak sidelobes behave as secondary main lobes. A strong target in the direction of a particularly large peak sidelobe is indistinguishible from the return through the main lobe and will introduce a false target on the display in the direction of the mainlobe. This effect is often evidenced on synthetic array radar maps where very strong targets (e.g., a radio tower) will show up at regularly spaced intervals corresponding to processor sidelobes.

The effect of integrated sidelobes is similar to that of "slot noise." For example, consider mapping a non-reflective area surrounded by a continuous field of reasonably bright reflectors. An example is a concrete or macadam road (which has a very small geometrical backscattering coefficient) surrounded by vegetation (which has an average backscattering coefficient). The return from the road must compete with the integral of the returns from all of the sidelobes of the vegetation. Hence, if a good integrated sidelobe power ratio is not obtained, small dark targets in a field of bright targets will be washed out.

We shall discuss the various, major sidelobe sources, one at a time in the following sections. These include motion compensation, storage media, filter formation, and transmitter/receiver stability. The antenna and atmosphere normally provide negligible contribution to these problems. The chapter concludes with examples of SAR imagery and future applications of SAR.

4.1 MOTION COMPENSATION

As discussed above, we have formed an effective antenna by moving the aircraft along a flight path which has been assumed to be straight. In real arrays, a reasonable criterion for straightness or flatness of the antenna surface is $\pm \lambda/4$, where λ is the radar wavelength. In synthetic array, due to the out-and-back nature of the radar, this tolerance must be halved. That is to say, we must either fly in a straight line to within $\lambda/8$, or we must correct for deviations from such a straight line motion. Considering the fact that $\lambda = 3.2$ cm at X-Band, and that the synthetic array length can be upwards of 100 meters long, it is not possible to control aircraft flight path this closely. In particular, it can be seen that the maximum uncompensated acceleration, a, acting for the time, T , to fly a synthetic array, must be related to the required accuracy by

$$\frac{1}{2}\, a \left(\frac{T}{2}\right)^2 \leqslant \frac{\lambda}{8} \tag{4.1}$$

or

$$a \leqslant \lambda/T^2 \tag{4.2}$$

Here, for an aircraft speed of 300 m/sec and an array length of 100 m, T will equal 100/300 = 1/3. For a λ of 3.2 cm, equation (4.2) will result in an acceleration tolerance of $a \leqslant 30$ cm/sec^2 $\cong$ 0.03 g.

To perform motion compensation, an acceleration can be measured along the line of sight between the aircraft and the region being mapped. Integrating such acceleration to obtain velocity and dividing by λ, yields the frequency change that must be applied to the doppler signals in order to correct for the measured motion. One means of introducing this frequency change is to vary the receiver local oscillator.

The accuracy of this process must be of the order of the maximum uncompensated acceleration, a, defined above. The process is further complicated by the fact that the line-of-sight accelerometer is also affected by gravity since the region being mapped lies below the aircraft. This "DC" acceleration must be subtracted in order to obtain the proper compensation.

4.2 STORAGE MEDIA

The phase effects of analog storage media are related to input-output processes. For photographic film or scan converter storage tubes, error contributors include non-linearities in the write and readout CRT sweeps (if used), non-linear film motion caused by reduction gear noise or film slippage, and improper compensation of the non-linear relationship between input signal amplitude versus output signal amplitude. A major advantage of digital processors is the elimination of this kind of phase distortion. In a digital processor, the only limiting factor is the linearity and dynamic range capability of the analog to digital (A/D) converter between the analog video signal and the digital storage media.

4.3 COMPLEX FILTERING

The complex filtering process discussed previously is necessary to separate individual targets at the same range and different azimuths. To obtain low peak and integrated sidelobes, it is necessary to provide suitable weighting of the received signals. Peak sidelobe levels of 25 to 30 dB can be achieved at the expense of a few percent spread in the 3 dB azimuth resolution. Figure 5-5 of Chapter V, and the discussion on sidelobe reduction in that chapter, illustrate what can be done with this process.

4.4 TRANSMITTER-RECEIVER STABILITY

A high degree of transmitter-receiver stability is required in order to maintain phase coherency for the entire duration of the synthetic array. For a maximum phase error of 90 degrees, for an X-Band carrier frequency of 10 GHz, and for a synthetic array time of one second, a stability of 1 in 4×10^{10} would be required. This accuracy is at the very limit of present state-of-the-art capabilities. The accuracy requirement can be greatly reduced by the use of the same oscillator both as the transmitted signal source and the local oscillator. The time for 90 degree stability is then reduced to 1 in 1.6×10^{7} for a 50 km maximum range. Such phase stability is readily achievable with temperature-controlled crystal oscillators. One satisfactory means of stabilizing all frequencies in a synthetic array radar system is to start with a single crystal at some convenient frequency such as 5 to 10 MHz and upconvert and downconvert to all of the other frequencies in the system including the PRF, the IF, the local oscillator frequency, and the transmitted frequency.

4.5 EXAMPLES OF SYNTHETIC ARRAY RADAR IMAGERY

The following example of synthetic array radar imagery is provided through the courtesy of the Jet Propulsion Laboratory, California Institute of Technology. The write-up is obtained from their publication entitled Seasat Log; August 21, 1978; No. 1.

The Seasat-A satellite system containing the synthetic array radar became operational on 26 June 1978. The semimajor axis of the satellite was 7170 km with a corresponding altitude of about 800 km. Specific characteristics of Seasat-A spaceborne SAR are listed in Table 4-1.

On 4 July 1978, synthetic array radar data was obtained during the Goldstone and Alaska pass and successfully processed. Subsequently, a long pass was obtained over Merrit Island Launch Area covering a swath from Colombia, South America, to the Canadian border. Ocean features observable in this data include internal waves, air-sea interactions marked by natural slicks, Gulf Stream shear similarly marked, orographic steering of on-shore winds by terrain features and surface waves, both swell (300 m) and locally wind generated (70 to 100 m). Beaufort Sea ice data was obtained on 11 July via Alaska, showing shore fast ice, the shore lead and sea ice off Banks Island. Hurricane FICO was observed on 15 July and storm seas of 150- and 250-m wavelength were clearly evident as was the confused sea near the eye of the hurricane. Thus, by mid-July, initial data on all of the principal synthetic array radar objectives had been successfully obtained and processed. Instrument and Data Processing System performance has been determined to be within specification in all tested categories.

TABLE 4-1 SEASAT-A SPACEBORNE SAR CHARACTERISTICS

PARAMETER	VALUE
RF Band	L
Frequency, MHz	1275
PRF, Hz	1555 ± 14%
Altitude, km	794
System Bandwidth, MHz	19
Peak Power, kw	0.8
Average Power, watts	40
Antenna Length, m	10.7
Antenna Width, m	2.16
Angle Off Vertical, degrees	20.5
Polarization	Horizontal
Single Look Resolution, m	7 X 25
Noise Equivalent, dB	–16.5
Integrated Sidelobe Level, dB	–8
Swath Width, km	100
Data Gathering Time, hours	300
Processing	Optical

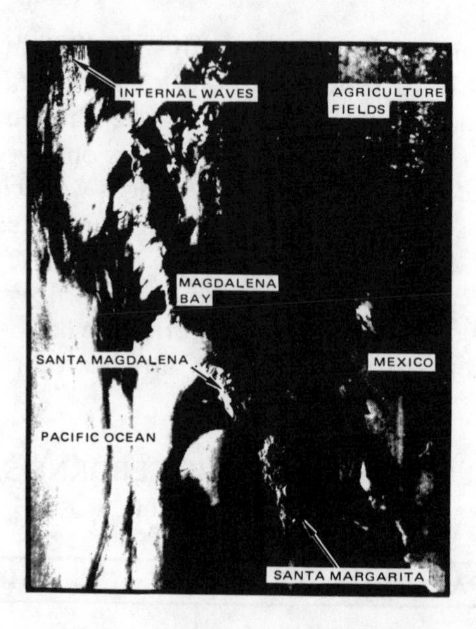

Figure 4-1 Synthetic Array Radar Picture of the Coast of Mexico Obtained via Satellite

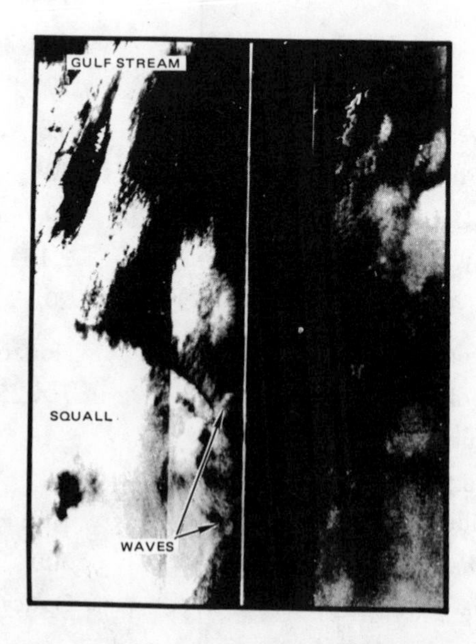

Figure 4-2 Synthetic Array Radar Picture of the Florida Coast Obtained via Satellite

Figures 4-1 and 4-2 provide two examples of synthetic array radar photographs obtained from the Seasat project.

Figure 4-1 shows a radar image of the Baja Peninsula coast of Mexico (right), a chain of coastal islands (center) and the Pacific Ocean (left) obtained by Seasat's Synthetic Aperture Radar prior to dawn on 7 July. The bright ocean areas show varied patterns including groups of internal waves and several areas of wind-roughened surface. (Corresponding weather maps indicated a 15- to 20-knot offshore breeze in these zones.) Dark patches in the lee of the mountinous islands are areas of water sheltered from the wind. North is toward the top of the "picture." Islands shown are Santa Margarita to the south and Santa Magdalena. Agricultural fields can be seen at upper right. Tidal channels connect with Almejas Bay (lower right) and Magdalena Bay (center of photo). This image is a segment of a SAR swath extending some 2500 miles along western North America to Alaska.

Figure 4-2 is an example of complex ocean structure which includes a portion of the Gulf Stream just off the Florida coast northeast of Miami and north of Grand Bahama Island in the Atlantic. The image covers an area approximately 60 by 75 miles (95 by 120 km). The pattern of striations at upper left may be related to the motion of the Gulf Stream current, while the irregular pattern in the lower left of the picture may indicate the effects of wind and rainfall on the ocean's surface resulting from a local rain squall. Ocean waves can be seen in a limited zone near the squall area (center). Although, at first glance, this photo appears to show clouds and other atmospheric features, the synthetic array radar sees through the atmosphere and records only reflections received from the surface. The image was acquired at 6:04 a.m. EDT on 8 July 1978.

4.5.1 Airborne SAR Imagery*

Figure 4-3 gives two examples of SAR imagery obtained by airborne radars. This figure shows a 20-foot resolution SAR image and an 80-foot resolution SAR image together with corresponding

---

*Provided through the courtesy of the Hughes Aircraft Company, El Segundo, California.

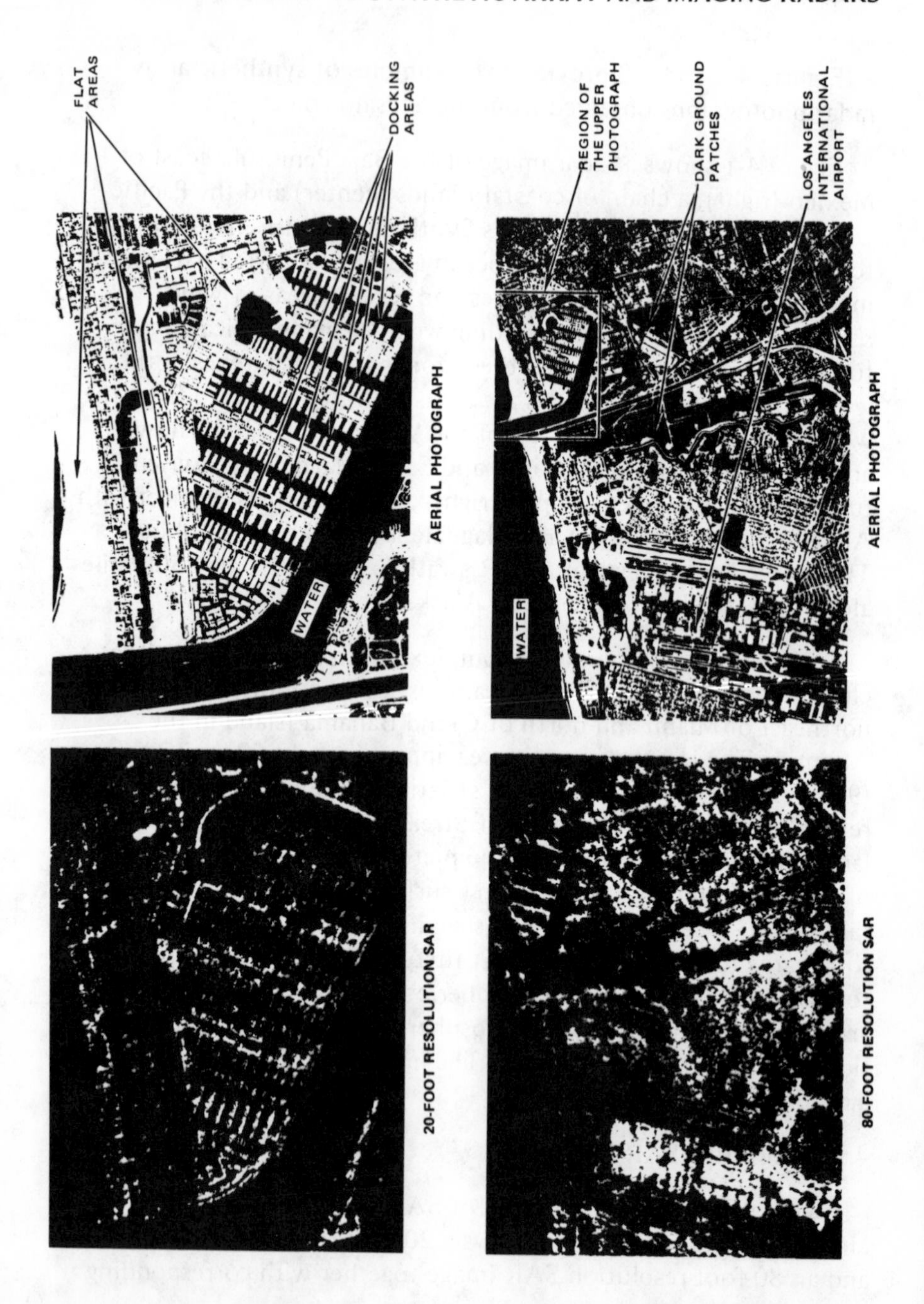

Figure 4-3 Synthetic Array Radar Imagery Obtained from Airborne SAR

aerial photographs in each case. These pictures represent the Marina del Rey area of the City of Los Angeles, California. Note that the upper photograph is a picture of the upper-right-side of the lower photograph. Water inlets and docking areas are clearly shown in the upper pictures and, to a lesser degree, in the lower pictures. These areas, because of low water backscattering coefficient, are shown as black areas in SAR pictures. Flat ground areas, also because of low backscattering coefficient, are shown as black areas in SAR images – making them indistinguishable from water covered regions.

An aerial photograph corresponding to the 20-foot resolution SAR image shows the white water wakes left by moving boats. These wakes, again because of low backscattering coefficient, are not shown in the corresponding SAR image. The moving boats, however, present a large radar cross section and should have been seen on the radar image. The reason for their disappearance is their motion which, because of range/range rate ambiguity inherent in SAR, places them in an incorrect range, getting lost in other radar returns. This range/range rate ambiguity will be discussed in Chapter V.

The bright spots on the left side of the 80-foot resolution SAR image correspond to large buildings and various structures in and around the Los Angeles International Airport. These buildings and structures, in effect, act as corner reflectors to incident radar energy giving rise to elevated target returns. These returns in turn show as bright spots on SAR images. The aircraft runways and roads, on the other hand, show up in SAR images as long black streaks because of their low backscattering coefficients.

Comparison of aerial and SAR photographs also reveals that dark ground patches on aerial photos do not necessarily show up as dark spots on SAR images, because these dark patches may have high backscattering coefficients which determine their brightness in SAR images. Differences between SAR and aerial photographs can be partially explained by noting that photographs are made from reflection of light in the visible frequency region of wavelengths 4 to 8 micron ($4\text{-}8 \times 10^{-6}$ cm) while radar images are made using radar wavelengths in the order of 3 cm (x-band). Since the reflectivity or backscattering is a function of wavelength, large differences between SAR and photographic images are to be expected.

4.6 DISCUSSION AND APPLICATION OF SYNTHETIC ARRAY RADARS

The following writeup on synthetic array radars is primarily obtained from a paper entitled "Future of Synthetic Aperture Radars," by Frank T. Barath and presented at the Electronics and Aerospace Systems Conference (EASCON '78), Arlington, Virginia, 24-27 September, 1978.

Airborne and spaceborne Synthetic Aperture Radar (SAR) is characterized by a number of important features:

- It is able to image a surface with very fine resolution of a few meters to coarse resolution of several kilometers.
- It can provide imagery to a given resolution independently of altitude, limited only by the transmitter power available.
- A number of fundamental parameters such as polarization and look angle can be varied to optimize the system for a specific application.
- Imaging is independent of solar illumination (availability or angle) because the system provides its own source of illumination.
- It can operate independently of weather conditions if sufficiently long wavelengths are chosen.
- It operates in a band of the electromagnetic spectrum different from the bands used by visible and infrared (IR) imagers.

Aside from the obvious advantage of being able to remotely sense an area in all-weather conditions, SARs in some applications yield a unique signature, and in other applications provide data complementary to those obtained from other sensors. An example of unique signature is in soil moisture determination: there is preliminary evidence of a peak of radar response to moisture in the top few feet of soil at a frequency of 4.7 GHz (6.5 cm wavelength). On the other hand, usefulness of SAR imagery as a complement to visible/IR data is exemplified by the study of geologic units through merged land and airborne SAR images.

TABLE 4-2 DEMONSTRATED SAR APPLICATIONS

APPLICATION	PHENOMENA
Water Resources	– Surface Water, Flood and Wetland Mapping – Drainage Basin Mapping – Lake Ice Mapping
Vegetation Resources	– Vegetation Type Determination
Geology	– Surface Structural Mapping – Geomorphology
Oceanology	– Directional Wave Spectrum Determination – Currents Boundary Detection – Ice Extent, Motion, Ridge and Lead Determination
Other	– Land Use Mapping – Oil Spill Detection – Ship Detection

4.6.1 State-of-the-Art SAR Applications

Table 4-2 lists those applications of SARs that are documented in the open literature and in general accepted by the remote sensing community. It should be kept in mind that the data base for developing these applications has been obtained by ground-based reflectometers and aircraft SARs only, and that considerable work is still underway in the relatively new area of remote sensing. What can be expected from spacecraft-borne SARs, therefore, is largely still an open issue. Three SAR system characteristics in particular cannot be adequately simulated from an aircraft: the swath width obtained from a spacecraft with a near-constant angle of incidence is considerably wider; the reflecting Fresnel zones are considerably larger from spacecraft than from aircraft altitudes; and the length of time a target is observed by a spacecraft SAR is substantially longer than in the case of an airborne system. These factors are considered by many to constitute compelling arguments for spaceborne SARs. The first steps in this direction are represented by the Seasat-A system launched on 26 June 1978 and discussed above.

4.6.2 SAR Technology

Today's SAR technology pertinent to remote sensing is represented by a number of aircraft SARs, which mostly are military systems derivatives, and the Seasat-A spaceborne SAR, a basically new design. In addition, the Apollo Lunar Sounder Experiment (ALSE) has been developed and was flown in 1972 in lunar orbit for studying subsurface layering. Although ALSE is a dual-frequency synthetic array system, its wavelengths of 6 and 60 m (50 and 5 MHz) have been found to be too long for use in terrestrial applications.

Synthetic array radar systems can be divided into three major functional blocks: the antenna, the sensor, and the data subsystems. The airborne systems have relatively straightforward antennas, either of the reflector or the phased array type. Spaceborne SAR antennas, on the other hand, because of distances involved, need to be considerably larger, thereby creating packaging, thermal and electrical design challenges, as exemplified by the Seasat-A system. This antenna is a uniformly illuminated, horizontally polarized, planar phased array and measures 10.7 X 2.2 m in size. It is made up of 8 identical panels, which fold up accordion-fashion for launch.

The sensor (transmitter, receiver, oscillators and mixers) technology is quite mature as represented by the airborne systems. The sensor design of the Seasat-A system is a relatively straightforward adaptation of this technology with one exception: advantage has been taken of recent high frequency power transistors to build an all-solid state, high efficiency, high reliability power output stage.

The data systems of both the airborne and the spaceborne SARs are the least developed at the present time, principally because of the large data volume, and the extremely complex data manipulation required to form imagery from the raw sensor output. Most airborne systems now record their raw data optically, either on-board, or on the ground through a wideband data link. Seasat-A telemeters the raw data in an analog format which is recorded digitally on the ground, later to be translated into an optical record. On-board digital recording systems and various types of digital processors are being developed at a rapid pace to overcome the shortcomings of analog/optical/film data systems. Hybrid digital ground processors (general purpose computers augmented with special purpose hardware such as array processors) are expected to be presently operational. Special purpose computers, the only variety capable of processing SAR data at the real-time rate, have so far been individually specialized to specific military aircraft radars and are difficult to convert to other systems. A major effort is underway by NASA to develop a flexible full-capability, real-time digital system, amenable to eventual use on-board spacecrafts.

4.6.3 The Future of SAR

Table 4-2 lists potential SAR remote sensing applications with only preliminary supportive evidence. The list is not necessarily complete inasmuch as new applications are being constantly investigated, and it is also to some extent artificial since the boundary between "proven" and "potential" is somewhat subjective.

The applications of SARs to the remote sensing of the earth's resources and environment are principally being developed by NASA under the Supporting Research and Technology (SR&T) program. An ambitious SR&T program has been proposed, aimed at developing the fundamental understanding of the surface-radar wave interactions, at designing the algorithms and techniques for extracting information from SAR imagery, and at exploring actual applications in the geology, hydrology, oceanology and vegetation resource areas. Additional research is proposed for studying such applications as fishing vessel monitoring, oil spill detection and urban morphology. Proposed spaceborne SAR systems include the Seasat follow-on SAR which is designed for the ocean-related

data flow logistics problems, and the SIR-B/C which is a project aimed at establishing a shuttle-based, versatile, reusable research facility capable of supporting a wide spectrum of applications through the late 1980's.

Near term technology improvements point to fully digital operation, including data link, recording, and processing. The rates are on the order of 60-120 MBps, and the processing is basically in real-time with possible ground-based special purpose correlators.

The longer-term (post-1985) technology requirements are far more challenging as some of the applications mature and enter operational or quasioperational phases. It is clear that the antenna systems need to be far more sophisticated to provide wide swath coverage and multiple frequency operation. High performance scanning-beam, multibeam, multipolarization antennas will have to be developed, tested, qualified for space. Phased arrays of various types are likely candidates, but lower cost and higher efficiency mechanically scanned reflector antennas might present attractive alternatives. Sensors will be pressed to be more efficient and provide more output power: higher frequency solid state transmitters, perhaps of the distributed kind, and low-noise integrated multichannel receivers will be highly desirable. Excellent reliability for long-term (3 years) in-space operation will be demanded. Finally, complete on-board real-time data correlation and processing to extract patterns/features for specific applications will be mandatory to stem the flood of output data and enhance the usefulness of the sensor to the ultimate user.

5 Pulse Compression Techniques and SAR Mechanization*

The average transmitted power of a given radar may be raised by increasing the pulselength within the given transmitter constraints. However, the increased pulselength (reduced receiver bandwidth) has the undesirable effect of reducing the range resolution capability of the radar. Knowing this, it is desirable to raise the transmitted power by increasing the pulselengths and simultaneously keeping a constant bandwidth. To achieve this, a long pulse containing some sort of phase or frequency modulation is transmitted. Upon reception, the pulse must be compressed to permit separation of adjacent range resolution cells as shown in Figure 5-1. Here, two received expanded pulses overlap but, after compression, they are time-separated and can be resolved on a display, exactly as if a short pulse had been transmitted.

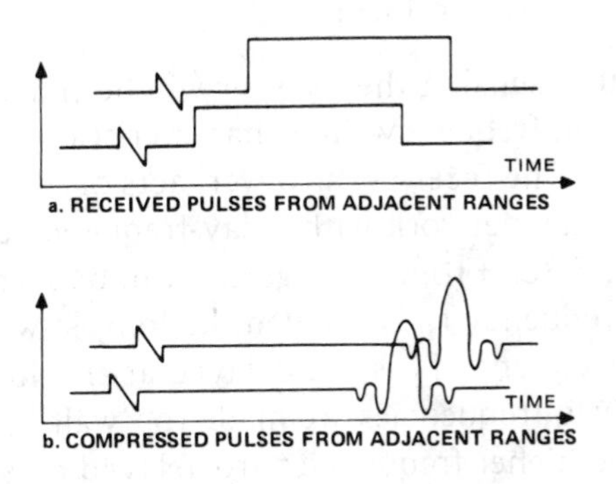

Figure 5-1 Transmitted and Received Pulses

*References 1, 3, 5 and 10.

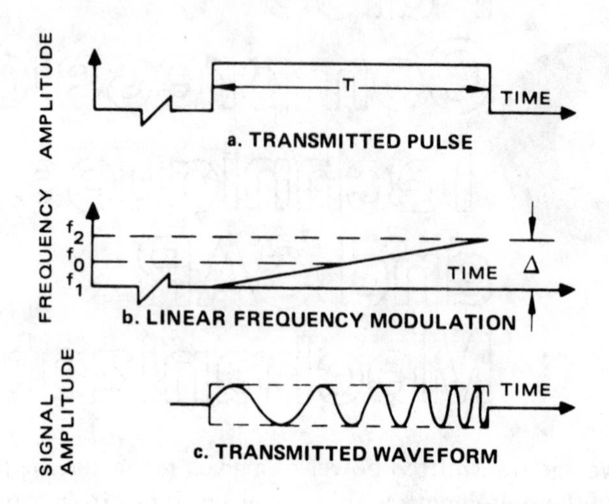

Figure 5-2 Transmitted Waveform of a Linear FM Pulse

Historically, the first such technique reduced to practice was "chirp."* Chirp achieves increased transmitter power and constant bandwidth by a linear frequency modulation of the transmitted waveform and a receiver delay network utilizing frequency-time characteristics of the transmitted waveform.

The frequency time characteristics of the transmitted signal are given in Figure 5-2. The transmitted pulse is of duration T and is frequency-modulated from f_1 to f_2 at a rate of Δ/T where $\Delta = f_2 - f_1$. The effect of frequency modulation on the transmitted sinusoidal signal is shown in Figure 5-2c.

The target return signal at the receiver will be similar to the transmitted signal in frequency-time characteristics as shown in Figure 5-3a and b. The receiver, however, acts on the received signal by a time delay network with delay-frequency characteristics as given in Figure 5-3c. From this figure it can be seen that the lowest received frequency f_1 is delayed the longest while the highest received frequency f_2 is not delayed at all. In this manner the initially received frequencies are made to "wait" while subsequently received higher frequencies are delayed by shorter amounts.

*Reference 10.

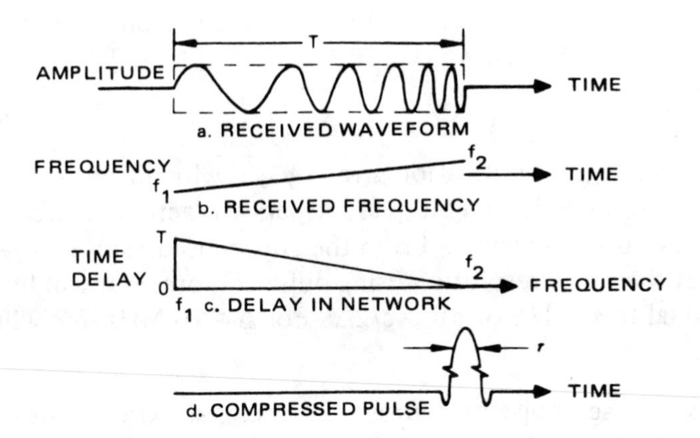

Figure 5-3 Received Waveform of the FM Pulse and Subsequent Pulse Compression

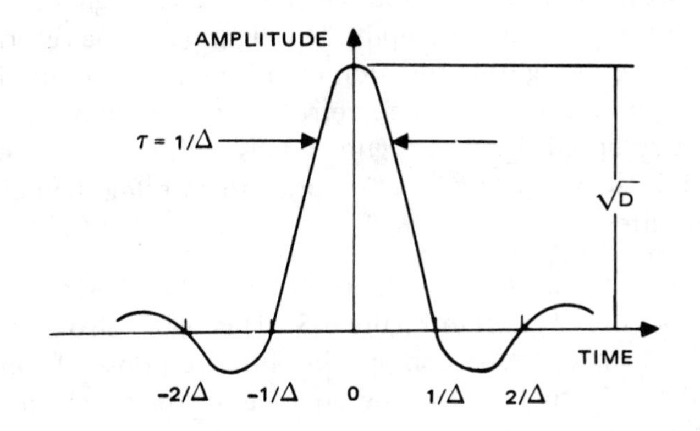

Figure 5-4 Amplitude-Time Characteristics of the Compressed Pulse

The final output, by conservation of energy, is a short pulse of large amplitude as given descriptively in Figure 5-3d. The amplitude-time characteristics of this pulse are given again in greater detail in Figure 5-4. In this figure the amplitude of the pulse is given relative to a maximum value of $\sqrt{D}$ where D is defined as the *dispersion factor* and is equal to $T\Delta$. The duration of the compressed pulse τ is proportional to $1/\Delta$ as seen from Figure 5-4. From these observations, we get

$$T/\tau \cong T\Delta = D \quad , \tag{5.1}$$

with signal voltage amplification given by $\sqrt{D}$. In current radars using pulse compression techniques, dispersion factors in excess of 1000 have been achieved. From the above discussion, it is noted that the range resolution using pulse compression will be proportional to $\tau = 1/\Delta$ or $\delta R = c/2\Delta$. For $\Delta = 10$ MHz, δR will be 15 meters.

The compressed pulse of Figure 5-4 displays sin x/x amplitude characteristics in much the same way as the antenna electromagnetic gain patterns. The sidelobes of the amplitude-time characteristics of Figure 5-4 are undesirable as they may give rise to false detections if time discrimination is used to calculate range. The peak and integrated average sidelobes of the received signal can be reduced by the application of amplitude weighting to the returned signal. A common weighting function used for sidelobe reduction is the various powers of the trigonometric cosine function. Effectively, the returned signal of Figure 5-3a is multiplied by the proper form of the weighting function prior to entering the pulse compression filter.

The output of the pulse compression filter for two cosine weighting functions is given in Figure 5-5. This figure also gives the unweighted sin x/x pulse compression filter response. From Figure 5-5 the weighting reduces the sidelobes of the filter while increasing the width of the mainlobe.*

5.1 MATHEMATICAL DERIVATIONS

In this section, mathematical derivations will be given for the justification of the above-stated facts. Since the convolution integral is used in these derivations, the characteristics of this integral will be briefly reviewed prior to mathematical derivations.

*Reference 7.

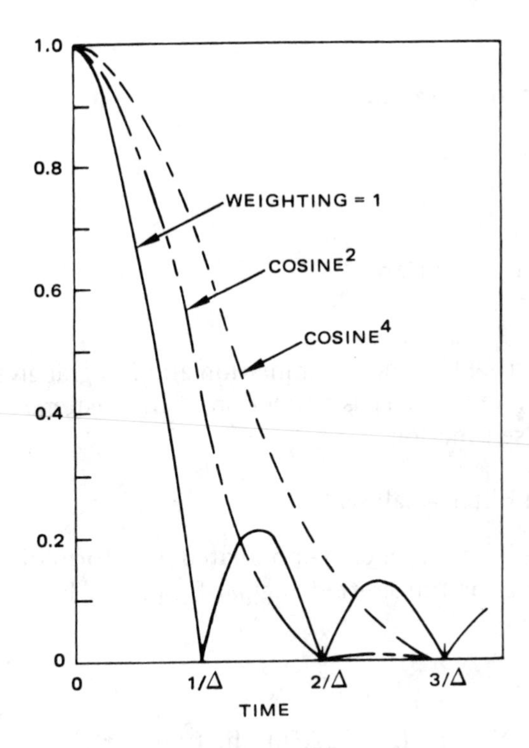

Figure 5-5 Sidelobe Suppression of Linear FM Compressed Pulse at the Matched Filter Output

5.1.1 Convolution Integral

Consider a system having a response, h(t), to a unit impulse function, $\delta(t)$, of unit area. It is desired to obtain the response of the system to a forcing function, f(t), in terms of h(t). In these expressions t represents time. The forcing function, f(t), can be imagined to consist of a number of impulse functions of duration, $\Delta\tau$, and amplitudes, $f(\tau)$. Assuming the magnitude of each impulse equals its area, the response of a system to an impulse at time, τ, will be

$$f(\tau)\Delta\tau h(t-\tau) \qquad (5.2)$$

The response of the system, g(t), to the forcing function, f(t), will be equal to the summation of impulse responses of equation (5.2) from t_1 to t_2 as given by

$$g(t) = \sum_{\tau=t_1}^{t_2} f(\tau)\, h\,(t-\tau)\, \Delta\tau \quad , \tag{5.3}$$

which goes to

$$g(t) = \int_{t_1}^{t_2} f(\tau)\, h\,(t-\tau)\, d\tau \quad , \tag{5.4}$$

as $\Delta\tau$ goes to zero. The above summation and integral gives the response of a system to $f(t)$ as a function of the system response to a unit impulse function.

5.1.2 Matched Filter Analysis

Consider the linear frequency-modulated waveform of Figure 5-2. Designating the transmitted angular frequency by

$$\omega_o + \frac{1}{2}\,\mu t \quad , \tag{5.5}$$

where μ is the FM slope ($\mu = 2\pi\Delta/T$), the transmitted signal can be written

$$f(t) = \cos\left(\omega_o t + \frac{1}{2}\,\mu t^2\right), \; -\frac{T}{2} \leqslant t \leqslant \frac{T}{2} \tag{5.6}$$

$$= 0, \quad \text{elsewhere.}$$

The matched filter has an impulse response, $h(t)$, that is time inverse of the signal at the receiver input as given below

$$h(t) = \sqrt{\frac{2\mu}{\pi}}\; \cos\left(\omega_o t - \frac{1}{2}\,\mu t^2\right), \; -\frac{T}{2} \leqslant t \leqslant \frac{T}{2} \quad , \tag{5.7}$$

where the $\sqrt{2\mu/\pi}$ is a factor that results in unity gain.

The return signal from the target at time t_o will have the characteristics of the transmitted signal, equation (5.6), except that it will be shifted in frequency proportional to the doppler shift, ω_d, as follows

$$f(t_o) = \cos\left[(\omega_o + \omega_d)\, t_o + \frac{1}{2}\mu t_o^2\right], \tag{5.8}$$

where now $f(t_o)$ represents the signal that is returned from the target. From this signal, the matched filter characteristics of equation (5.7), and the convolution integral of (5.4), we get the general output of the matched filter as

$$g(t_o, \omega_d) = \sqrt{\frac{2\mu}{\pi}} \int_{-T/2}^{T/2} \cos\left[(\omega_o + \omega_d)\tau + \frac{1}{2}\mu\tau^2\right] \cos\left[\omega_o(t_o - \tau) - \frac{\mu}{2}(t_o - \tau)^2\right] d\tau \tag{5.9}$$

The above integral can be evaluated for specific values of the parameters using numerical integration methods and digital computers. This evaluation will result in values of $g(t_o, \omega_d)$ as a function of time, t_o, and doppler shift, ω_d. The closed form solution of equation (5.9) can also be obtained through a considerable amount of trigonometric and algebraic manipulation. The result of this calculation is

$$g(t_o, \omega_d) = \sqrt{\frac{\mu}{2\pi}} \cos\left[\left(\omega_o + \frac{\omega_d}{2}\right) t_o\right] \times \frac{\sin\left[\frac{\omega_d + \mu t_o}{2}(T - |t_o|)\right]}{\frac{\omega_d + \mu t_o}{2}}, \quad -\frac{T}{2} \leqslant t \leqslant \frac{T}{2} \tag{5.10}$$

$|t_o|$ = absolute value of t_o

The above expression represents the output of the pulse compression matched filter as a function of time and the doppler shift of the received signal. Note the sin x/x function appearing in (5.10). Several additional points about the above equation need further discussion.

It might appear from the form of equation (5.10) that $g(t_o, \omega_d)$ has been doppler shifted $\omega_d/2$ instead of the usual amount of ω_d. However, expansion of the cosine and sine terms yield terms of the type

$$\cos\frac{\omega_d t_o}{2}\sin\frac{\omega_d t_o}{2} = 1/2 \sin \omega_d t_o\ , \tag{5.11}$$

and no terms involving $\omega_d/2$ only. (This is as would be expected from a passive, linear filter.) Equation (5.10) for cases with no doppler shift ($\omega_d = 0$) becomes

$$g(t_o) = \sqrt{\frac{\mu}{2\pi}}\ \cos\omega_o t_o\ \frac{\sin\left(\frac{\mu t_o}{2}\right)(T - |t_o|)}{\frac{\mu t_o}{2}} \tag{5.12}$$

To a good approximation, the maximum value of $g(t_o, \omega_d)$ given by equation (5.10) occurs at $t_o \ll T$ and $\omega_d + \mu t_o = 0$ resulting in

$$t_o = -\frac{\omega_d}{\mu} = -\frac{Tf_d}{\Delta}\ , \tag{5.13}$$

where f_d is the doppler frequency shift and Δ is the frequency excursion of the transmitted frequency modulated signal. Since the values of Δ and T are known, from (5.13), the time at which the maximum amplitude occurs will depend on the doppler shift of the returned signal. In radars using time discrimination ranging, the peak value of the returned signal determines the transmit-receive time delay that determines radar-target range. The effect of doppler shift on this timing will result in ambiguities that, in effect, trade range information for closing-rate information. An intuitive discussion of range/range rate ambiguity will be given in the next section.

By studying equations (5.10) and (5.12), note that the (sin x)/x function is only the multiplier of the harmonic cosine function appearing in these equations. This harmonic function contains the basic frequency ω_o plus the doppler shift information which, over several pulses, will be utilized to obtain doppler or azimuth resolution information. As will be seen in Chapter VI, a single returned pulse does not provide a sufficient number of samples for doppler shift determination. For this reason, several returned pulses are needed.

5.2 RANGE/RANGE RATE AMBIGUITY

In radars using time discrimination ranging, the peak value of the returned signal determines the transmit/receive time delay which in turn determines radar-target range. When the target is moving, its doppler shift may affect the time of the arrival of the peak signal causing an ambiguity which trades range information for closing rate or range rate information. Figure 5-6 gives a more vivid picture of amplitude-frequency-timing relationship in pulse compression radars. From this figure it is noted that closing targets with a positive doppler shift ($f_d > 0$) will appear closer while opening targets with a negative doppler shift ($f_d < 0$) will appear farther away. Nonmoving targets with zero doppler shift ($f_d = 0$) will appear at the correct range.

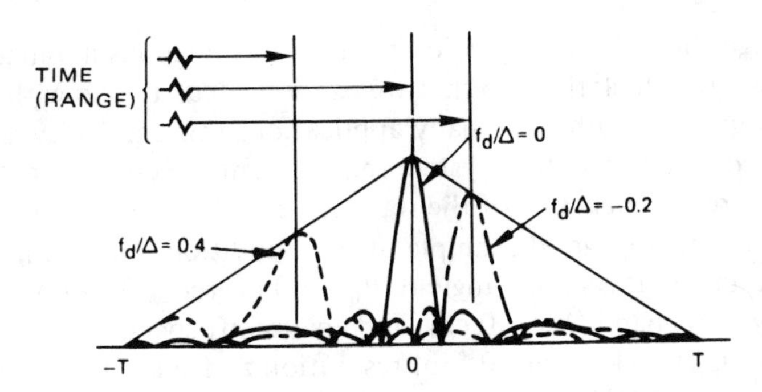

Figure 5-6 Amplitude-Frequency-Time Relationship of a Pulse Compression Signal

To fully understand the shift in time scale due to target motion shown in Figure 5-6, consider the following discussion. Figure 5-7 gives the time frequency relationship of a frequency modulated signal. This figure also gives target return signal from a non-moving, a closing, and an opening target. The non-moving target return is only delayed by the round-trip travel time of electromagnetic radiation. The closing target return signal, in addition to this delay, is doppler-shifted by the amount f_d due to the radar-target closing rate (Figure 5-7b). Remembering the frequency-time characteristics of the pulse compression filter, the "time in filter" for this signal will be shorter than that of the non-moving target by t_o as shown in the figure.

Similarly, for an opening target, Figure 5-7c, the filter response will be delayed by t_o as the filter's triggering frequency is delayed. With the above discussion, it is easy to derive the expression for t_o as

$$t_o = -\frac{Tf_d}{\Delta} \tag{5.14}$$

For closing targets with f_d positive the peak signal will be reached sooner than for a non-moving target. For opening targets with f_d negative the peak signal will be reached later than for a non-moving target. Both conditions were shown previously in Figure 5-6.

Although the pulse compression radars are ambiguous in range and range rate calculations so that we cannot calculate one without a knowledge of the other, in many applications this ambiguity is tolerable compared to other system requirements. Synthetic array radars, where doppler shift of the returned signal is minimal compared to range, are a good example of this condition. For example, consider a target at a squint angle of $\theta_0 = 60$ degree. The doppler frequency, at X-band ($\lambda = 3.0$ cm) and an aircraft speed of $V = 300$ m/sec, is 10 kHz. For 50 foot resolution and small angles of incidence, $\Delta = 10$ MHz and $f_d/\Delta = 10^{-3}$ – corresponding to a negligible ambiguity level.

For other combinations of the above parameters the doppler shift might have a pronounced effect on radar-target range. This frequency/range ambiguity, together with the fact that doppler shift is used for azimuth resolution, may give rise to mispositioned moving targets in radar ground maps. This may be the case of trains showing up off their tracks or cars off the highways.

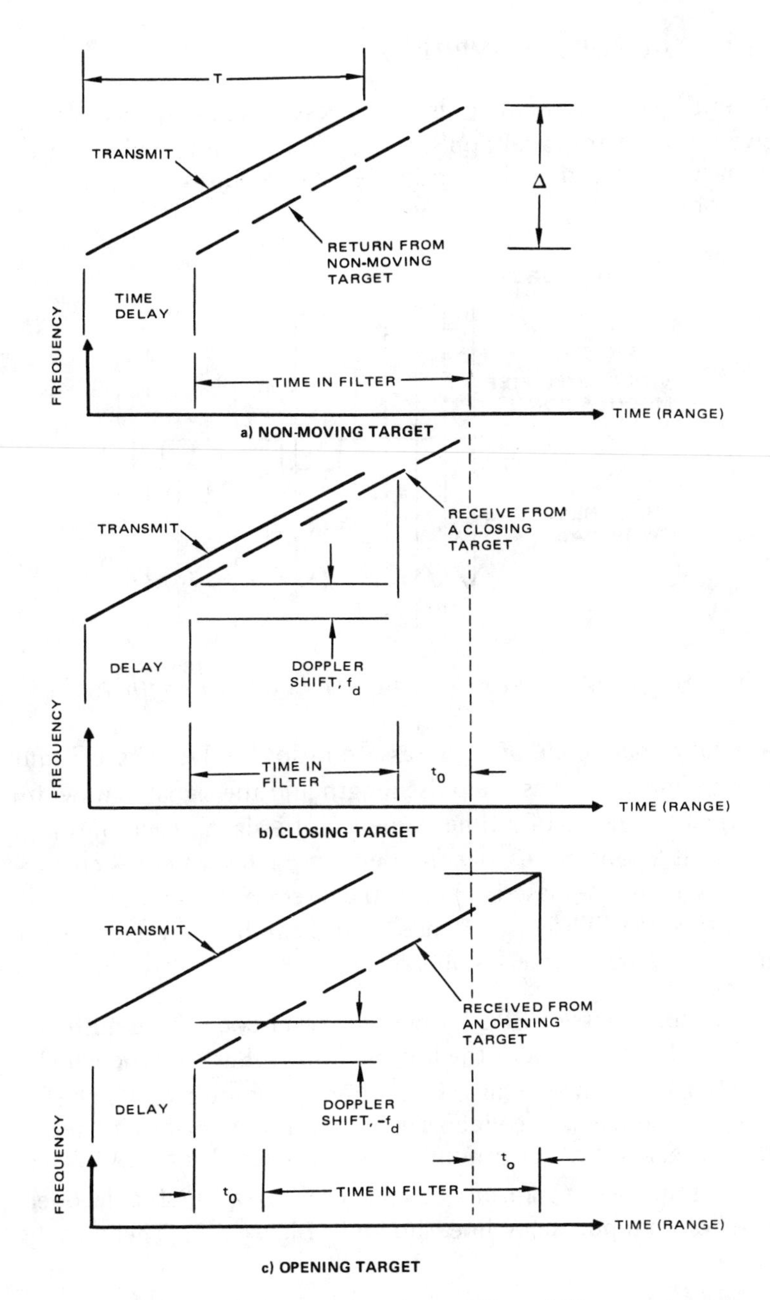

Figure 5-7 Time in Pulse Compression Filter Before and After Peak Value of Non-Moving Target Returns

5.3 DIGITAL PULSE COMPRESSION

Recently, various digital pulse compression techniques have received interest for radar applications. Two such techniques are the binary phase code (BPC) and poly-phase codes such as the Frank code.*

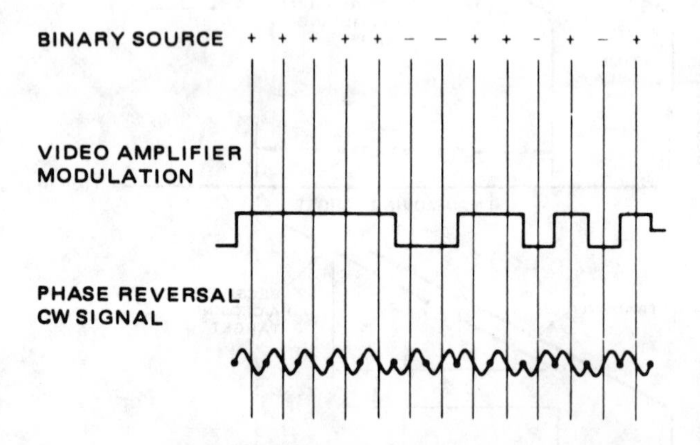

Figure 5-8 Waveforms for a Barker Code of Length 13

A binary phase code of compression ratio, $\xi = T\Delta$, where T and Δ were defined above as the pulse length and the signal bandwidth, respectively, consists of a time sequence of ξ signals each of duration $1/\Delta$, at a constant carrier frequency. Each segment is characterized by being in-phase (+1) or 180 degrees out-of-phase (−1) relative to some CW signal. Hence, a BPC can be represented by a sequence of ξ symbols, each plus or minus.

A particular class of BPC known as Barker codes have quite low sidelobes. Let us consider the longest known Barker code which contains 13 elements. This is illustrated in Figure 5-8, which shows the binary sequence, the video signal as a frequency-modulated squarewave, and the corresponding phase reversals of a CW signal. If such a sequence is transmitted and received, a suitable decoder consists of a tapped delay line, shown in Figure 5-9. There, each

*Reference 3.

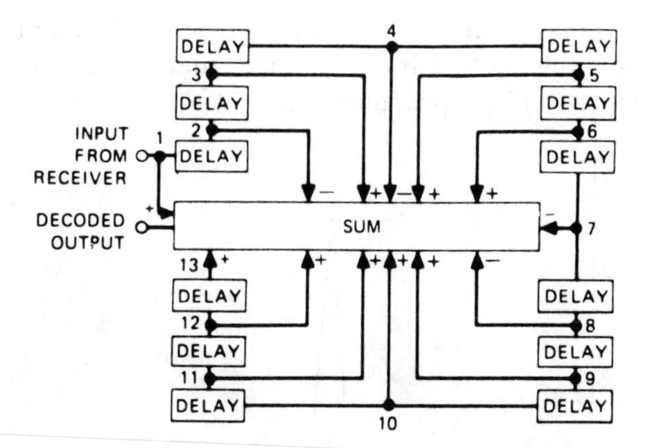

Figure 5-9 Barker Decoder of Length 13

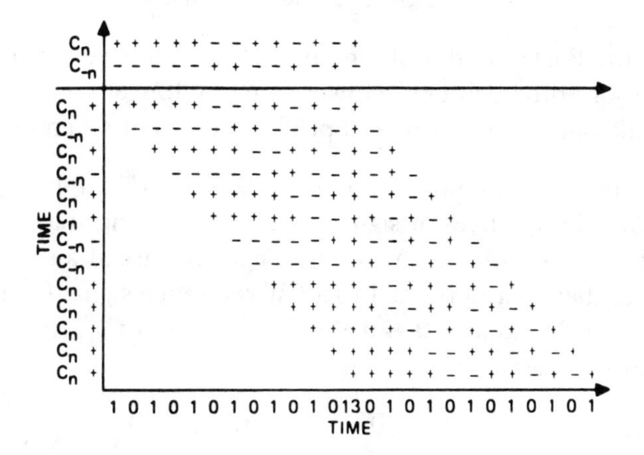

Figure 5-10 Procedure for Calculating the Decoder Output for Barker Code Containing 13 Elements

of the 13 taps adds or subtracts according to the code. The total delay is T and the individual delays are $1/\Delta$, the length of a segment. Figure 5-10 gives the procedure for calculating the decoder output and Figure 5-11 gives a plot of the output. The input signal appears in the decoder for 26 segment times from first appearance at tap 1 to last appearance at tap 13.

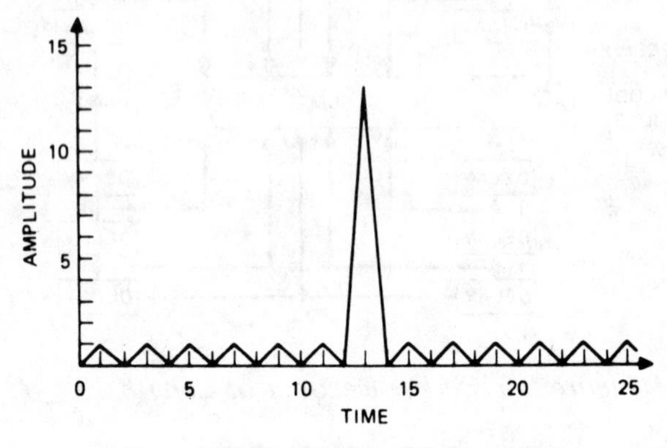

Figure 5-11 Barker 13 Decoded

Only the Barker codes of lengths 13 and less have the unique property of unit sidelobes. However, many longer codes exist with peak values up to 150 and peak sidelobes of seven or less.*

A Frank code of compression ratio, $\xi = N^2$, (N an integer) consists of N subsequences of signals, each subsequence having N segments $1/\Delta$ seconds long. Again, the segments are at a constant carrier frequency and related to a CW reference signal by a phase angle of $2\pi n/N$ radians, $0 \leqslant n \leqslant N\text{-}1$. In terms of N, the generalized Frank code can be written

$$\begin{array}{ccccc} 0 & 0 & 0 & 0 \;\ldots\; & 0 \\ 0 & 1 & 2 & 3 \;\ldots\; & N\text{-}1 \\ 0 & 2 & 4 & 6 \;\ldots\; & 2(N\text{-}1) \\ & & & \text{o} & \\ & & & \text{o} & \\ & & & \text{o} & \\ & & & \text{o} & \\ 0 & (N\text{-}1) & 2(N\text{-}1) & 3(N\text{-}1) \;..\; & (N\text{-}1)^2 \end{array} , \qquad (5.15)$$

*Reference 3.

where each row in the matrix is a subsequence of N elements with linearly increasing phase. Since frequency is the rate of change of phase with respect to time, linearly increasing phase implies constant frequency. Thus, each row presents a constant frequency.

As we proceed from one row to the next the rate of change of phase of the new row is more rapid than that of the previous row, i.e., 0, 2, 4, 6 instead of 0, 1, 2, 3. This means that the new row represents a higher frequency than the previous row. With this explanation it is easy to see that the Frank code is a stepwise approximation to the swept frequency employed in "chirp". As a result, the same sidelobe level and weighting considerations apply for both Frank code and Chirp pulse compression.

For an unweighted Frank code of length 16, the phase angles and the corresponding complex numbers are

$$\begin{matrix} 0 & 0 & 0 & 0 \\ 0 & 90 & 180 & 270 \\ 0 & 180 & 0 & 180 \\ 0 & 270 & 180 & 90 \end{matrix} \qquad \begin{matrix} 1 & 1 & 1 & 1 \\ 1 & j & -1 & -j \\ 1 & -1 & 1 & -1 \\ 1 & -j & -1 & j \end{matrix} \tag{5.16}$$

The correlation function is calculated as shown in Figure 5-12 and its magnitude is plotted in Figure 5-13. The decoder mechanizations would be generally the same as in Figure 5-9 except that the 16 required inputs to the summer would be weighted by the complex conjugate of the corresponding Frank code term as shown in Figure 5-12.

Digital codes, such as Barker and Frank codes described above, are transmitted using one basic frequency and coding the signal for transmission by shifting its phases as shown in Figure 5-8. The received signal from each coded pulse will also contain the phasing of the transmitted signal. In addition, if one considers the returns of several transmitted pulses from a single point on the ground, one can detect the quadratic phase shift between these pulses as discussed previously and shown in Figure 2-7. Similar to the previous discussion, this phase shift is used in obtaining azimuth or y-direction resolution. The x-direction resolution is obtained from the measurement of the arrival time of the decoded pulse.

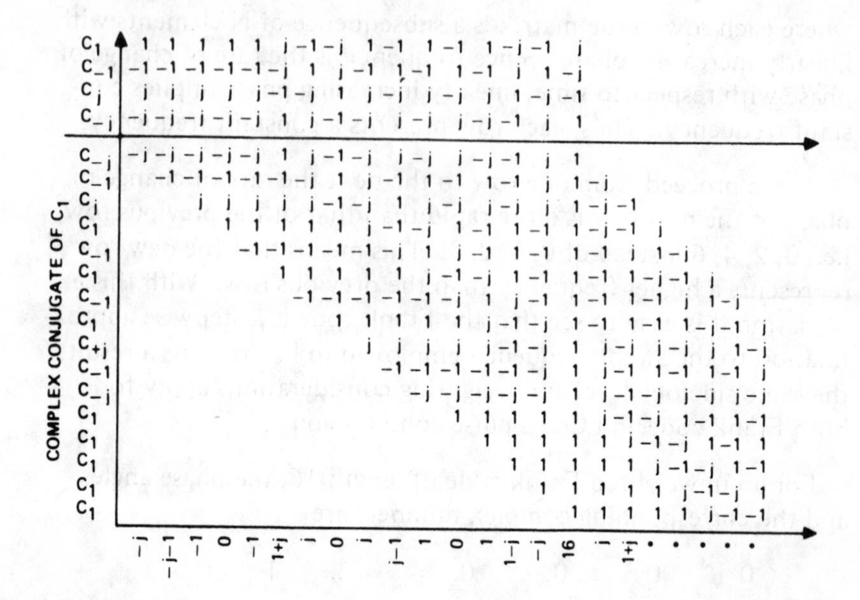

Figure 5-12 Calculation of the Correlation Function of Frank Code 16 Elements

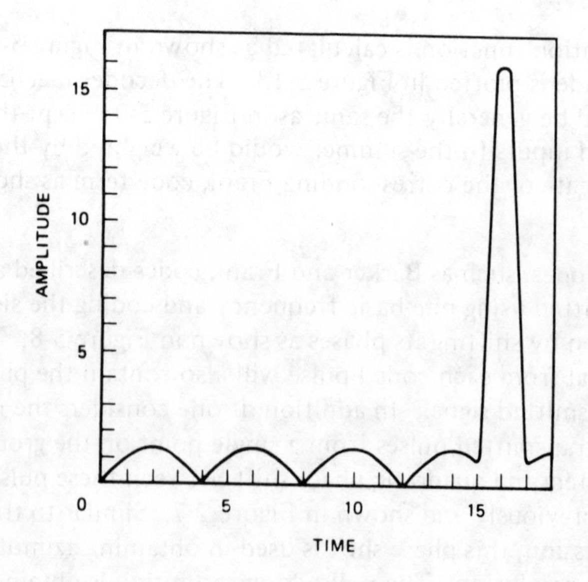

Figure 5-13 Frank Code of 16 Elements

It has been shown that by linear operations on the received signal the effective pulsewidth can be considerably reduced. This allows the transmission of longer duration pulses without the adverse effects of increased pulse length on target resolution capability. With pulse compression techniques, overlapping target return pulses can be separated in time at the output of pulse compression filter. Sidelobe level can be reduced by amplitude weighting in the compression filter. Time delay-doppler ambiguity of pulse compression radar is an important consideration in the design and application of high range resolution radars.*

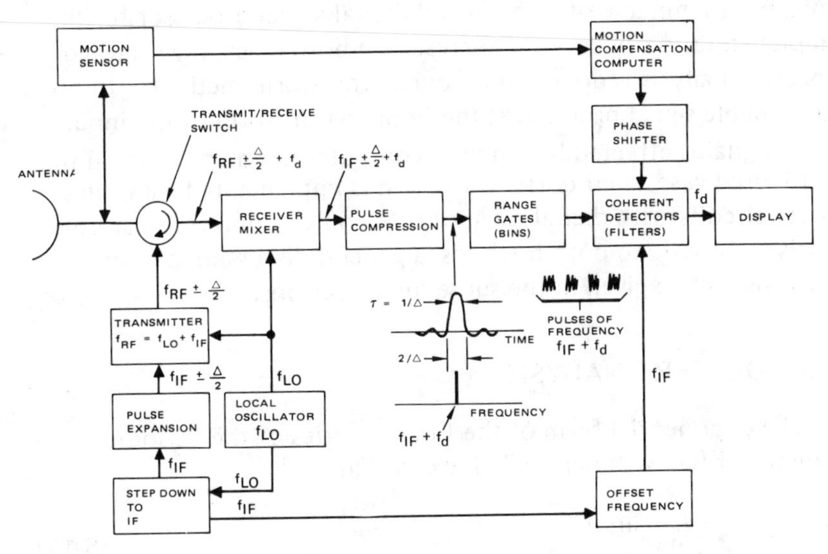

Figure 5-14 SAR Mechanization Block Diagram

5.4 MECHANIZATION BLOCK DIAGRAM

Figure 5-14 gives the mechanization block diagram of a SAR using pulse compression techniques. The stable local oscillator provides the basic frequency to the transmitter and acts as the reference frequency in determining the IF system frequency. The transmit/receive switch changes the radar function between transmission and reception of radar signals. Motion sensors measure

*Reference 14.

motion of the antenna as input to a motion compensation computer for subsequent phase correction and focusing calculations. The coherent detectors take frequency information from each range bin and extract corresponding doppler or azimuth resolution information. Note that this doppler information is the result of several returned pulses from the target (see next chapter). The range or x-direction resolution is obtained from the range gates of the figure.

5.4.1 Digital Mechanization

The coherent detectors (filters) of Figure 5-14 can be replaced by a programmable or a hardwired digital signal processor for the implementation of spectral analysis. In both cases, the required spectral analysis is done using Fourier transform methods. In programmable signal processors, the Fourier transform of the input time signal is effected by a digital computer program, while in the hardwired case, the Fourier transform is implemented using electrical circuits with digital components. Since Fourier transform analysis is basic to both methods, a general discussion of this analysis will be given in the subsequent sections.

5.5 FOURIER ANALYSIS*

The exponential form of the Fourier series of the periodic function F(t), with period T, is expressed as

$$F(t) = \sum_{n=-\infty}^{\infty} a_n e^{jn\omega t} \tag{5.17}$$

with

$$a_n = \frac{1}{2} \int_{-T/2}^{T/2} F(t) e^{-jn\omega t} dt, \tag{5.18}$$

where

*References 6, 7, and also 5 Chapter 3.

ω = angular frequency = $2\pi f$

f = frequency = $\frac{1}{T}$

T = period

j = $\sqrt{-1}$

In equations (5.17) and (5.18) it is assumed that the function is periodic with period T and that the coefficient a_n represents the integral of the function over a complete period. The corresponding Fourier transform pair of the function F(t) is expressed

$$F(t) = \int_{-\infty}^{\infty} G(\omega)e^{j\omega t}d\omega \quad (5.19)$$

and

$$G(\omega) = \frac{1}{2\pi}\int_{-\infty}^{\infty} F(t)e^{-j\omega t}dt, \quad (5.20)$$

where $G(\omega)$ is the exponential Fourier transform of F(t). Equation (5.20) of the Fourier transform can be compared to equation (5.18) of the Fourier series if we assume the period of the function F(t) to extend over the entire time domain $-\infty < t < \infty$. In practice, however, the function F(t) is given, starting from time zero and extending over a span of time, T, as given in Figure 5-15a. This will set the limits of equation (5.20) between zero and T. Given the Fourier transform $G(\omega)$ of the function F(t), equation (5.19), referred to as the inversion integral, will result in the function F(t). For the discrete representation of equations (5.19) and (5.20), consider Figure 5.15. The time span, T, of Figure 5-15a is divided into K equal increments of Δt each. The angular frequency span ω_N is divided into N equal increments of $\Delta\omega$ each. The frequency interval, $\Delta\omega$, is defined as

$$\Delta\omega = 2\pi\,\Delta f = 2\pi\frac{1}{T\cdot\frac{N}{K}} = \frac{2\pi K}{NT} \quad (5.21)$$

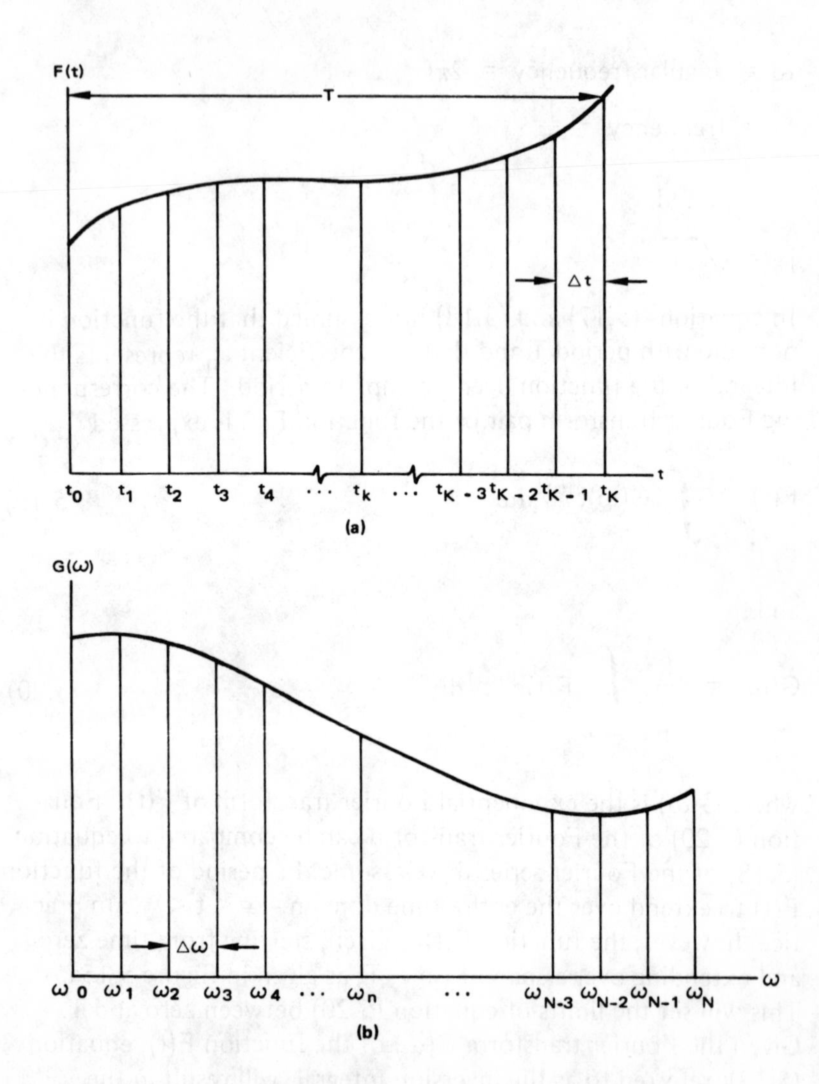

Figure 5-15 Discrete Representation of F(t) and G(ω)

In the above equation, the value of Δf is set at K/NT in order to simplify the method of calculation. Setting the number of time and frequency increments equal, N = K, which is usually the case in calculations, will result in a Δf of 1/T. Note that filter frequency bandwidth or filter spacing Δf is equal to 1/T where T is the data acquisition period or coherent integration time. Thus, there is a one-to-one correspondence between Fourier analysis and the analog processing discussed before. Equations (5.19) and (5.20) with the above notation can be written in discrete form as follows

$$F(t_k) = \Delta\omega \sum_{n=0}^{N-1} G(\omega_n)e^{j\omega_n t_k} \tag{5.22}$$

$$G(\omega_n) = \frac{\Delta t}{2\pi} \sum_{k=0}^{K-1} F(t_k)e^{-j\omega_n t_k} \ , \tag{5.23}$$

where the limits of the first sum, instead of -N/2 to N/2 (corresponding to equation (5.19)), are set from 0 to N-1 for computational convenience. This changes the foldover frequency from zero to $\omega_{N/2}$ as will be discussed later. The limits of equation (5.23) are taken to represent the available range of the variable F(t). Note that equations (5.22) and (5.23) represent the areas under the curves of Figure 5-15. Using the following definitions

$$\begin{aligned} F_k &= \frac{T}{2\pi K} F(t_k) \\ G_n &= G(\omega_n) \\ \omega_n &= n\Delta\omega = \frac{2\pi nK}{NT} \\ t_k &= k\Delta t = \frac{kT}{K} \ , \end{aligned} \tag{5.24}$$

equations (5.22) and (5.23) can be rewritten

$$F_k = \frac{1}{N} \sum_{n=0}^{N-1} G_n e^{(2\pi j/N)(nk)} \tag{5.25}$$

$$G_n = \sum_{k=0}^{K-1} F_k e^{(-2\pi j/N)(nk)} \tag{5.26}$$

Denoting $W = \exp(-2\pi j/N)$, equation (5.26) can be written in matrix form as

$$\underset{(N\text{x}1)}{\left[G_n\right]} = \underset{(N\text{x}K)}{\left[W^{(nk)}\right]} \underset{(K\text{x}1)}{\left[F_k\right]} \quad \begin{matrix} n=0, 1, \ldots, N\text{-}1 \\ k=0, 1, 2, \ldots, K\text{-}1, \end{matrix} \tag{5.27}$$

where subscripts n and k represent the row number of column matrices G and F, respectively. The product nk, in addition to being the power of W, represents the row, n, and column, k, in locating the W^{nk} element of the W matrix.

Equations (5.26) and (5.27) are the basic equations used in the development of computational logic of the Fourier transform. Based on these equations, several computer algorithms, under the general heading of Fast Fourier Transform, are currently available for the evaluation of spectral contents G_n of the time function F_k. These algorithms are used in programmable signal processors as software packages.

Equations (5.25) and (5.26) can also be mechanized by using hardwired digital logic. In this case, the coherent detectors (filters) of the analogue system are replaced by equivalent digital filters as discussed below.

5.6 QUADRATURE COMPONENTS

In digital signal processing, it is desirable to represent signals in quadrature components (i.e., in phase and 90 degrees out of phase). Consider Figure 5-16, with the input signal given below

$$S_1 = A \cos \left[(\omega_o + \omega_1)\, t + \phi_o\right] \tag{5.28}$$

The quadrature components of this signal are given as follows

$$
\begin{aligned}
S_2 &= A\left\{\cos\left[(2\omega_o + \omega_1)\,t + \phi_o\right] + \cos(\omega_1 t + \phi_o)\right\} \\
S_3 &= A\left\{-\sin\left[(2\omega_o + \omega_1)\,t + \phi_o\right] + \sin(\omega_1 t + \phi_o)\right\} \\
S_4 &= A\cos(\omega_1 t + \phi_o) \\
S_5 &= A\sin(\omega_1 t + \phi_o)\ ,
\end{aligned}
\tag{5.29}
$$

where the values of S_4 and S_5 after analog/digital (A/D) conversion represent the in-phase and 90-degree out-of-phase components x and y, respectively. Note that the low pass filter of Figure 5-16 eliminates the high frequency components of S_2 and S_3. The quadrature components x and y are sampled at intervals of Δt.

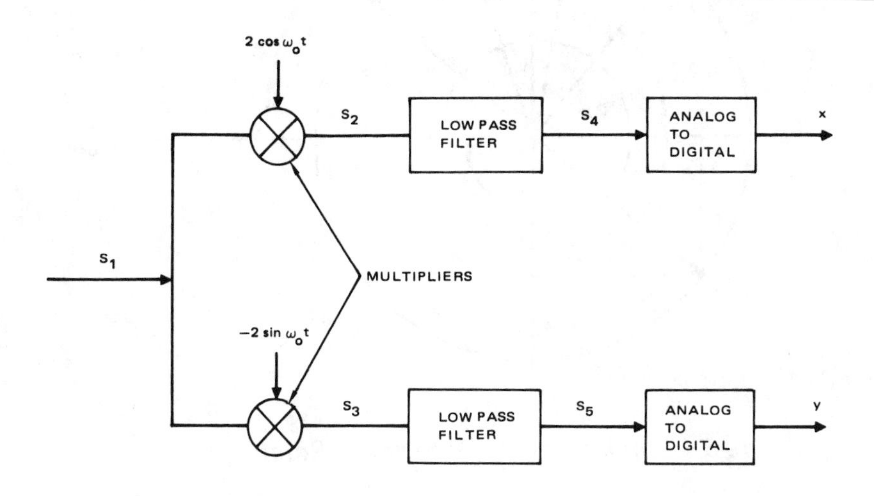

Figure 5-16 Quadrature Components of Input Signal

The following relations apply to the sampled values

$$
\begin{aligned}
x_k &= A\cos(\omega_1 t_k + \phi_o) \\
y_k &= A\sin(\omega_1 t_k + \phi_o) \\
|A_k| &= A,\ \phi_k = \omega_1 t_k + \phi_o\ ,
\end{aligned}
\tag{5.30}
$$

where, according to equation (5.24), t_k represents the successive sampling times. Thus, the sampled values of the quadrature components of input signal may be considered as a series of rotating phasors which rotate with an angular velocity ω_1 starting from the angular position ϕ_o, as shown in Figure 5-17.

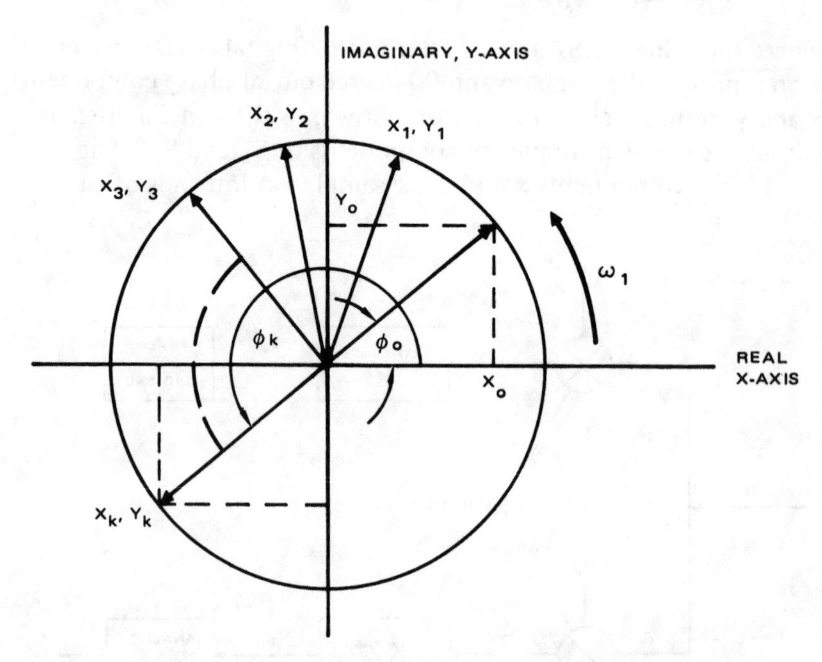

Figure 5-17 Input Signal as a Series of Rotating Phasors

5.7 DIGITAL MECHANIZATION OF THE FOURIER TRANSFORM

The Fourier transform of the time-domain function f(t) can be evaluated using several available algorithms written for digital computers. These algorithms basically evaluate relation (5.26) for G_n, given the input signal F_k. The evaluation of G_n can also be accomplished using digital filters where digital components and arithmetical operations are involved. In this case, the input signal F_k is represented by its quadrature components x_k and y_k, i.e.,

$$F_k = x_k + jy_k \tag{5.31}$$

Using this in equation (5.26) and substituting trigonometric functions for the exponential function, we get

$$G_n = \sum_{k=0}^{K-1} (x_k + jy_k)\left(\cos\frac{2\pi n}{N}k - j\sin\frac{2\pi n}{N}k\right) \tag{5.32}$$

Denoting $\frac{2\pi n}{N} = n\psi$ where $\psi = \frac{2\pi}{N}$, equation (5.32) becomes

$$G_n = \sum_{k=0}^{K-1}\left[x_k\cos(n\psi)k + y_k\sin(n\psi)k\right] + j\left[y_k\cos(n\psi)k - x_k\sin(n\psi)k\right] \tag{5.33}$$

Figure 5-18 shows the digital mechanization of equation (5.33). The circuitry of this figure uses multipliers, summers, delay networks, and gates. The sampled values of x_k and y_k are entered and

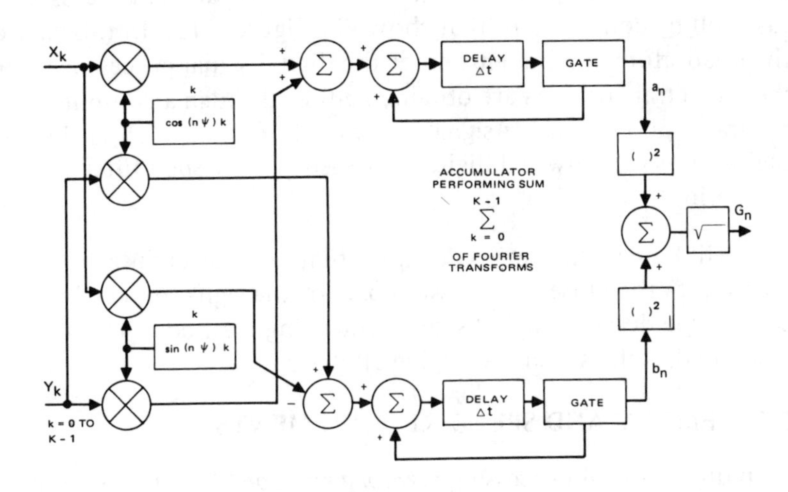

Figure 5-18 Digital Filter for the Computation of G_n

multiplied by appropriate values of sines and cosines. These values are summed in the accumulators, using delay networks with Δt time delay. The gates of the summing networks release the summation after K steps. The quadrature components a_n and b_n are then squared and the square root obtained to form the absolute value of G_n

$$G_n = \sqrt{a_n^2 + b_n^2} \tag{5.34}$$

Note that G_n represents the amplitude of the signal having a frequency of $n\Delta\omega$ (see equation (5.24)). Thus, for each frequency $n\Delta\omega$ where n=0, 1, 2, . . . , N-1 there should exist a digital filter similar to that of Figure 5-18. This set of N digital filters will detect signals present in the input (S_1 of Figure 5-16) having frequencies from 0 to $N\Delta\omega$ at steps $\Delta\omega$. This is similar to the operation of a series of tuned analog filters placed for the detection of signals.

5.8 DIGITAL SIGNAL PROCESSING BLOCK DIAGRAM

With the above discussions it is easy to see that the SAR mechanization block diagram of Figure 5-14 for the digital processing case will be converted to that shown in Figure 5-19. In this figure the quadrature components enter the digital signal processor where their spectral contents are obtained either through a computer program (programmable signal processor) or by a set of hardwired digital filters (hardwired digital processor). One such filter was shown in Figure 5-18.

It will be seen in the next chapter that using stretch radar techniques the pulse compression part of the signal processing can also be done by digital signal processing methods. This effectively makes the signal processing all digital.

5.9 MEMORY AND SPEED REQUIREMENTS*

In digital signal processing *memory* and *speed* are the determining factors of signal processing requirements. Equation (3.12) of Chapter III gives the number of samples as

---

*Reference 9.

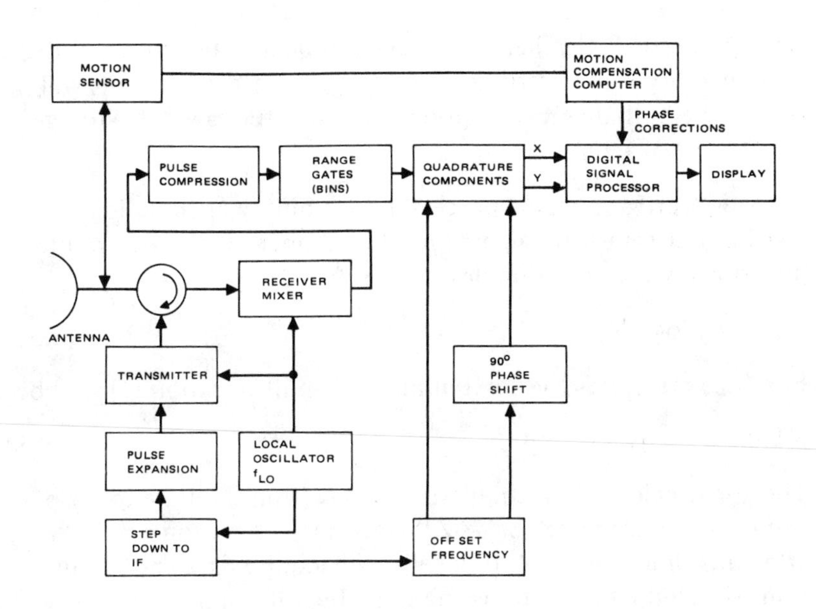

Figure 5-19 SAR Mechanization Block Diagram with Digital Signal Processor

$$
\begin{aligned}
N &= \frac{\lambda R}{(\delta AZ)^2} N_R \\
&= N_D N_R \ ,
\end{aligned}
\tag{5.35}
$$

where N_R is the number of range gates, N_D is the number of azimuth samples, and δAZ is substituted for azimuth resolution D/2. The number of samples N also determines the bulk memory (BM) requirements of the digital computer. Note the dependence of N on wavelength λ, range R, and the square of azimuth resolution δAZ.

Note that the actual raw data rate into the data processing system is determined by the value of pulse repetition frequency f_r used by the radar. That is, the number of samples coming into the system per second will be equal to f_r. If the number of samples per second needed for spectrum analysis

$$f_s = N_D/T \tag{5.36}$$

is less than the PRF, then a prefilter can be inserted ahead of the bulk memory. The prefilter is sampled at a rate of f_s. This will reduce the bulk memory requirements over the raw data storage by a factor of f_s/f_r.

Using a programmable processor with binary arithmetic and a fast Fourier transform algorithm, the number of complex multiplications NC for N_D samples will be**

$$NC = N_D \log_2 N_D \tag{5.37}$$

For N_R range gates the total number of multiplication NT will be

$$NT = N_R \cdot N_D \log_2 N_D \tag{5.38}$$

The above calculations should be made within the time to fly one azimuth resolution cell or $\delta AZ/V$ seconds, the reason being that after this time new resolution cells will acquire data and calculations should be repeated. Using this, the arithmetic rate NTR will be

$$\begin{aligned} NTR &= N_R (N_D \log_2 N_D)/(\delta AZ/V) \\ &= \frac{V N_R N_D \log_2 N_D}{\delta AZ} \end{aligned} \tag{5.39}$$

The above rate (complex multiplications per second) is made practical by use of parallel multipliers in a high speed digital computer. Note that the multiplication rate given by (5.39) is used for real time calculations where ground maps are simultaneously displayed. In a number of cases synthetic array radar data is collected for off-line calculations, in which case, the calculation rate and data acquisition rate need not match.

Since most computers operate with binary arithmetic, the number of bits in each memory location and thereby the total number of memory bits, can be computed from required signal-to-noise ratio coverage. For example, a S/N power ratio of 40 dB will require

**Reference 6.

$$10 \log \frac{S}{N} = 40 \text{ dB}$$

$$\left(\frac{S}{N}\right)_{\text{power}} = 10^4$$

$$\left(\frac{S}{N}\right)_{\text{voltage}} = \pm 10^2 \text{ (RMS)}$$

$$= \pm\ 141.4 \text{ (peak-to-peak)}$$

$$= \pm 2^{7.14} \qquad (5.40)$$

This value puts a requirement of eight (seven plus sign) binary bits on each memory location as given by N of equation (5.35). Additionally, the arithmetic unit of the computer should perform required calculations with eight bit word lengths.

Using the previously discussed ideas, the number of samples can be written in terms of the number of filters as follows

$$N = \text{(no. of filters)(filter BW)(2)(filter TC)} \quad , \qquad (5.41)$$

where BW and TC refer to bandwidth and time constant, respectively, and 2 is the number of samples per cycle per second. Since (BW) (TC) = 1, equation (5.41) becomes

$$N = \text{(no. of filters)(2)} \qquad (5.42)$$

Thus, the number of samples is twice the number of filters or equivalently, the number of filters is one-half the number of samples.

It was stated that in using FFT algorithms the number of samples and the number of filters are set equal for computational expediency. This works out well, as one-half of the computational filters, because of frequency foldover, are redundant. That is, filters symmetrically located with respect to the foldover frequency contain the same signal amplitudes. Thus, by setting the number of filters and samples equal and using FFT algorithms, the number of filters with useful information will be that which is required by signal processing N_D.

6 Two-Dimensional Correlation—Design Examples

Figure 6.1 shows signal processing principles used in a pulse compression synthetic array radar.* The transmitted radar pulses consist of long pulses of duration T with linearly increasing transmitted frequency from f_1 to f_2. The total amount of frequency increase of each pulse is Δ. The returned signal from ground points

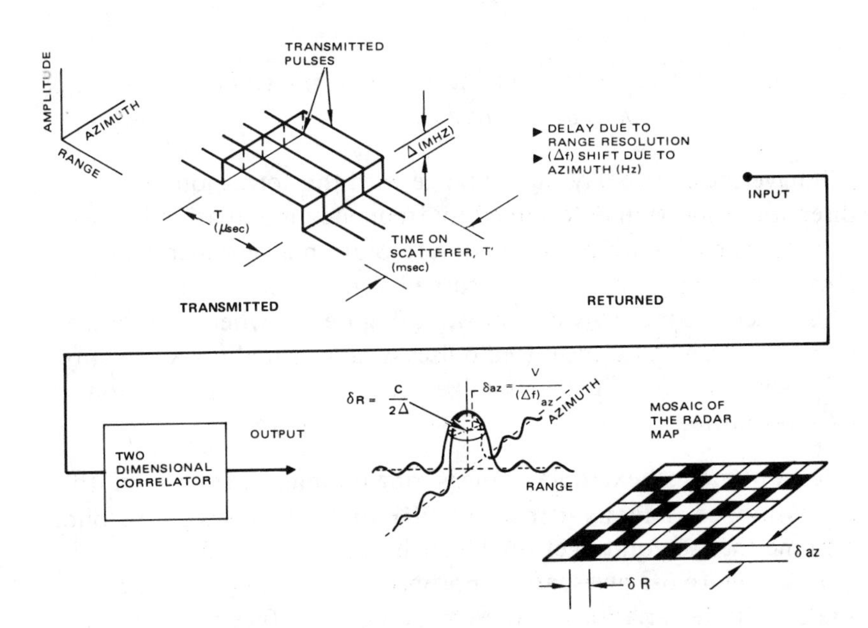

Figure 6-1 Two-Dimensional Synthetic Array Correlator

*Reference 12.

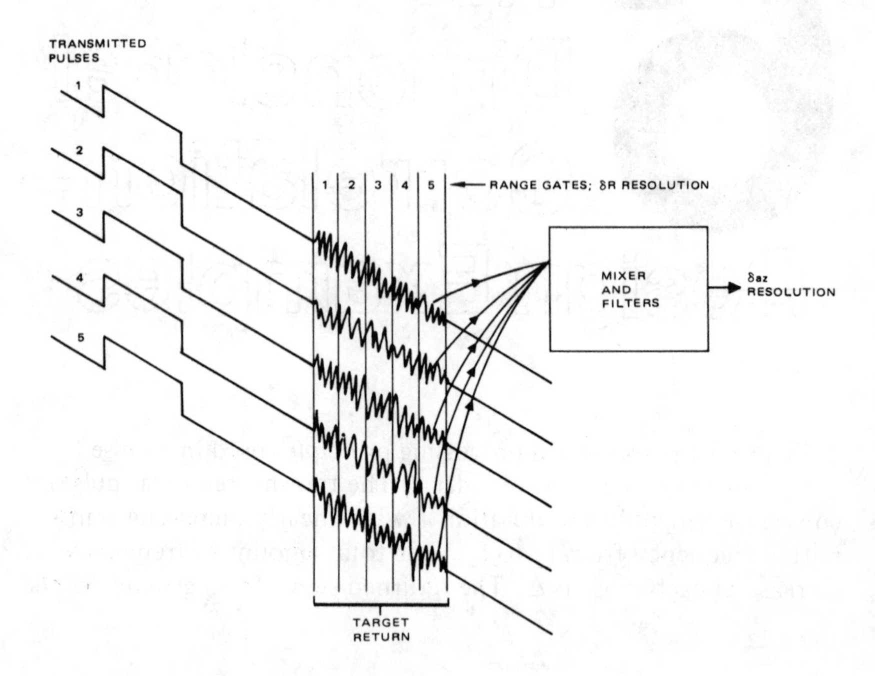

Figure 6-2 Returns from Several Transmitted Pulses are Range Gated and Filtered

will have the time delay due to range with the corresponding x-direction resolution determined by frequency modulation excursion Δ. The azimuth or y-direction resolution is obtained from measuring frequency shift of returned signals from several pulses. This process is more vividly shown in Figure 6-2 where the return signal from several transmitted pulses is range gated for x-direction δR resolution, and frequency mixed and filtered for y-direction or δAZ resolution.

As a numerical example of measuring doppler shift or azimuth resolution, consider a radar with a PRF of 1500 pulses per second. Assume that the on-target time is such as to result in 20 returned pulses. Figure 6-3 gives, after appropriate frequency reductions, the amplitude variations of these pulses for a 150 cycles per second doppler shift. It is noted that 150/1500 = 1/10 cycle per pulse. For 20 pulses the number of cycles will be *two* as shown in Figure

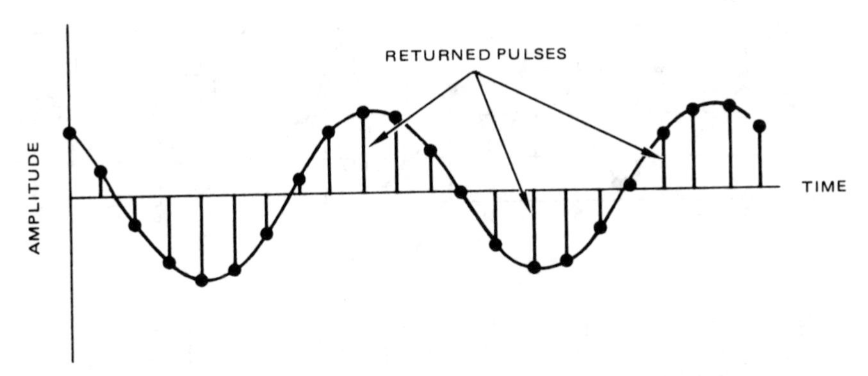

Figure 6-3 Doppler Shift Measurement from 20 Returned Pulses

6-3. Thus, if the sampled amplitudes shown by dots in Figure 6-3 were put through a 150 Hz tuned filter, the filter will respond with the appropriate sinewave amplitude shown in the figure. In a similar manner the presence of signal in several azimuth resolution cells can be detected by tuned filters or spectral analysis in the case of digital signal processing. Figure 6-4a represents a time signal containing frequencies of 300, 600, and 1050 Hz with respective amplitudes of 3, 5 and 2. If the pulse amplitudes shown by dots in Figure 6-4a are put through a set of tuned filters with center frequencies of 150, 300, 450, . . . the proper filters will respond as shown in Figure 6-4b.

Passing the return signal (Figure 6-1) through a two-dimensional correlator, we will get the output shown in the figure.

In the following discussion we will show the similarities between range and azimuth resolution cell behavior as depicted in the output of the correlator of Figure 6-1. Figure 6-5 shows the frequency-time relationship of the signal of a single frequency-modulated pulse of duration T. This pulse after the pulse compression filter will result in a compressed pulse duration of $\tau = 1/\Delta$ which corresponds to a two-way travel distance of

$$\delta R = c/2\Delta \quad , \tag{6.1}$$

where c is the velocity of propagation and δR is the range resolution.

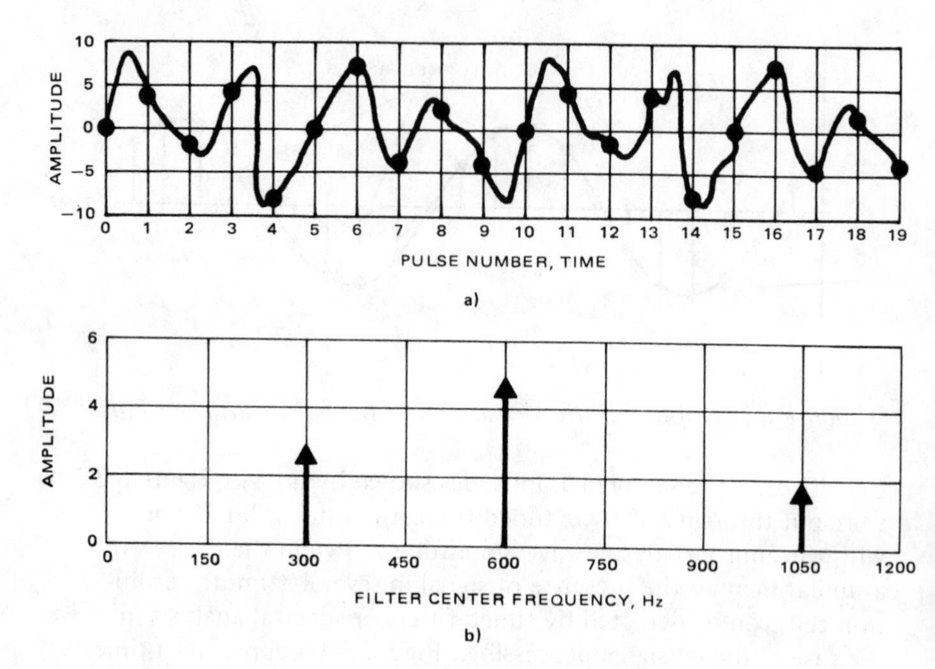

Figure 6-4 a) Presence of Multiple Frequencies in Time Signal, and b) its Resolution into Amplitude-Frequency

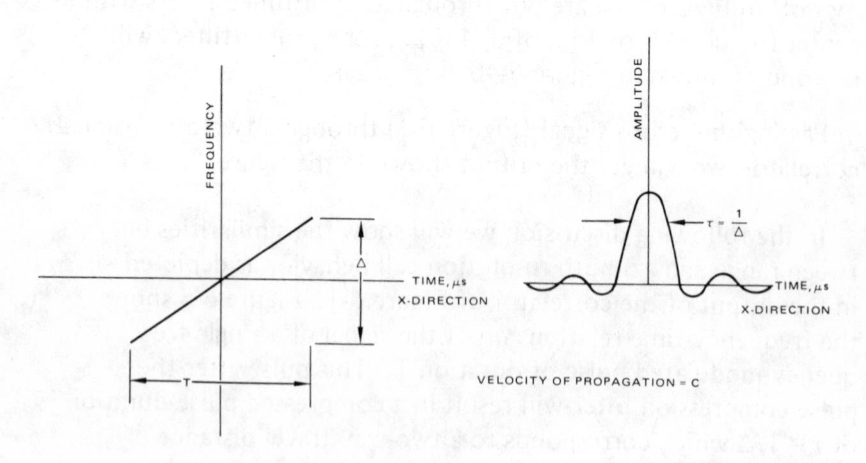

Figure 6-5 Range or X-Direction Resolution with Pulse Compression

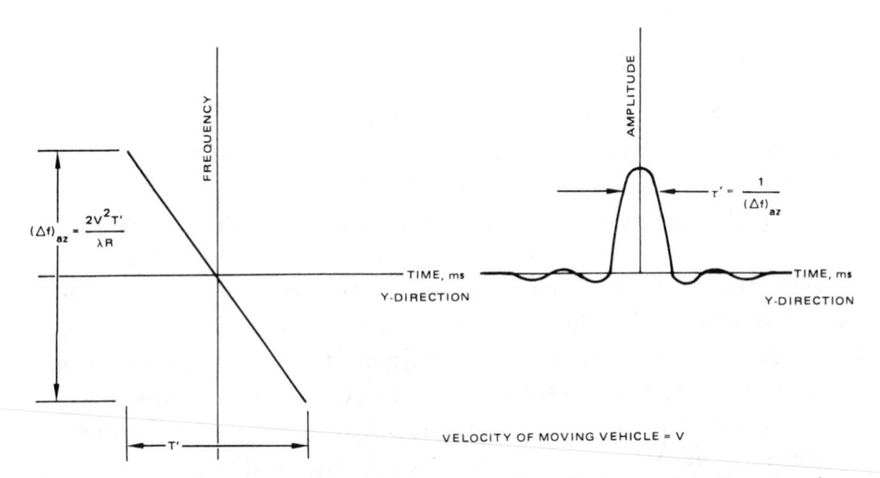

Figure 6-6 Azimuth or Y-Direction Resolution with Received Frequency Excursion

In a similar fashion, Figure 6-6 gives frequency-time relationship for azimuth resolution cells. Note that in this case time is measured in milliseconds while, for range resolution, it was measured in microseconds. Additionally, the azimuth frequency excursion of Figure 6-6 represents returns from several transmitted pulses but, for the purposes of discussion, this return is shown as a continuous function. In analogy to Figure 6-5, the corresponding time to azimuth resolution excursion of $(\Delta f)_{az}$ will be

$$\tau' = \frac{1}{(\Delta f)_{az}} , \tag{6.2}$$

with the same type of sin x/x amplitude-time behavior as before. Since in the azimuth direction the applicable velocity V is that of the moving vehicle (the aircraft or the spacecraft), the time τ' will correspond to a distance of

$$\begin{aligned} \delta AZ &= \tau' V \\ &= \frac{V}{(\Delta f)_{az}} , \end{aligned} \tag{6.3}$$

where δAZ is the azimuth resolution as shown in Figure 6-1. Figure 6-1 also shows a mosaic of the radar map with the above resolution elements.

6.1 STRETCH RADARS*

A somewhat newer method of range resolution by pulse compression is devised and used by stretch radars. This method obtains range or x-direction resolution through frequency analysis of returned data. A mechanization diagram of the radar is shown in Figure 6-7a. The duplexer of the figure switches the radar between transmit and receive functions. The frequency modulated transmitted signal is delayed and frequency shifted to be mixed with target return signal as shown in Figure 6-7b. Thus, x-direction delays specifying range resolution can be converted to frequency differences as shown. These frequency differences, in turn, can be computed using Fourier analysis. The advantage of the method lies in the fact that constant frequencies of Figure 6-7c allow the use of efficient fast Fourier transform algorithms and signal processing methods. The modulated frequencies of Figure 6-7b are not readily adaptable to digital signal processing methods. Additionally, the total frequency excursion shown by dots of Figure 6-7b, is considerably less than the frequency excursion of transmitted and received signals. This allows a lower number of samples to be used in representing the received signal for digital signal processing.

6.2 A DESIGN EXAMPLE

It is desired to design an airborne synthetic array radar to obtain a 3-by-3 meter resolution at the design point with aircraft flying at 250 m/sec at an altitude of 5000 meters. The radar-ground mapping range is to be 10 km.

The following additional radar parameters are assumed:

antenna beamwidth = 3.0 degrees

transmit frequency = 10 GHz

pulse duration = 1 μs

receiver bandwidth = 10^6 Hz

*Reference 1, page 269, contains the basic paper on stretch technique by W.J. Caputi.

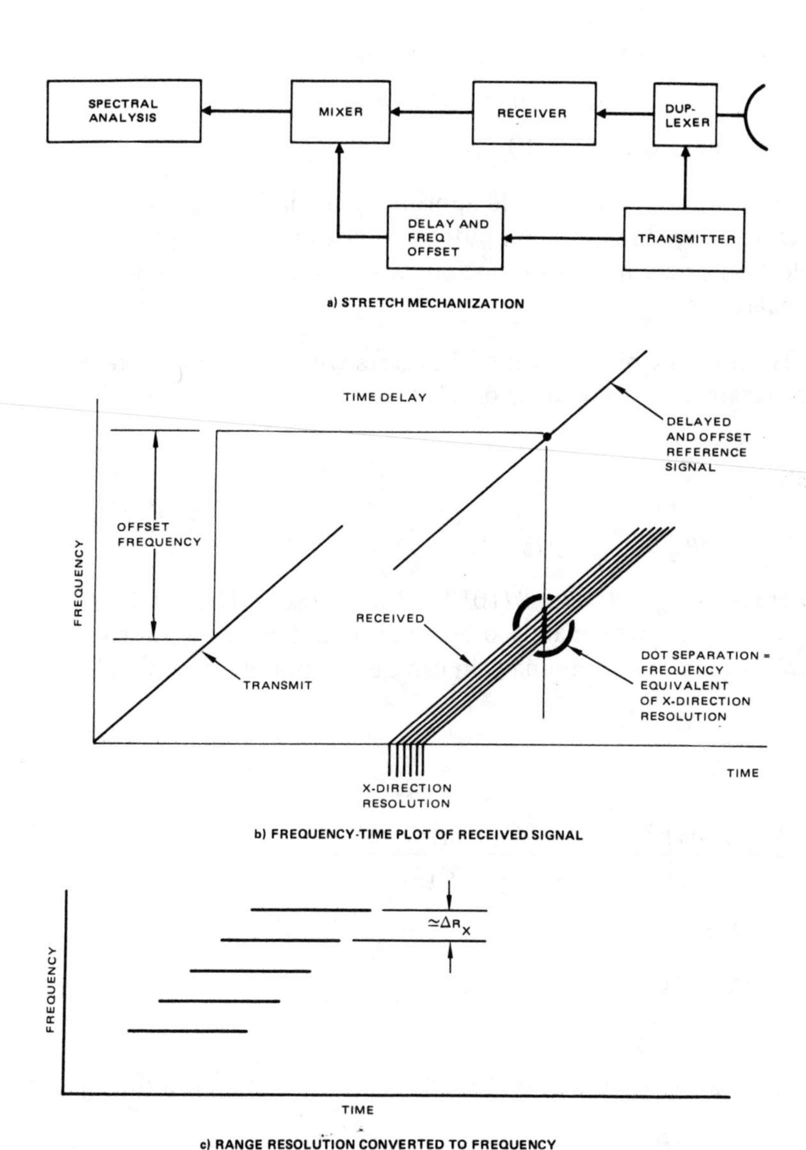

Figure 6-7 Compressed Pulse Signal Processing

pulse repetition frequency = 1500 Hz

equivalent noise temperature = 600° K

system losses = 10 dB (6.4)

With the above values and appropriate equations given in the text we can proceed to compute synthetic array radar parameters. Additional parameter assumptions will be made as needed in the calculations.

The antenna beamwidth of 3 degrees will result in an antenna size (diameter) from equation (1.9) of

$$\theta_{BW} = \frac{\lambda}{\ell}$$

$$\ell = \lambda/\theta_{BW} = 3/(3/57.3) = 57.3 \text{ cm} \tag{6.5}$$

where $\lambda = c/f_o = 3 \times 10^{10}/10^{10} = 3$ cm is used. The antenna diameter of approximately 0.60 meter is suitable for airborne applications. The antenna gain can be computed from (1.10) as

$$G = \frac{4\pi A}{\lambda^2}$$

$$= \frac{(4\pi)(\pi\ell^2/4)}{\lambda^2} = \frac{(4)(\pi)(\pi)(57.3)^2}{(4)(3)(3)}$$

$$= 3600$$

$$= 35.6 \text{ dB} \tag{6.6}$$

The value of $\tau = 1$ μs will result in an x-direction resolution of

$$\Delta R_x = \frac{c\tau}{2 \cos \alpha}$$

$$= \frac{3 \times 10^{10} \times 1 \times 10^{-6}}{2 \cos 30°}$$

$$= 173.2 \text{ meters} \quad , \tag{6.7}$$

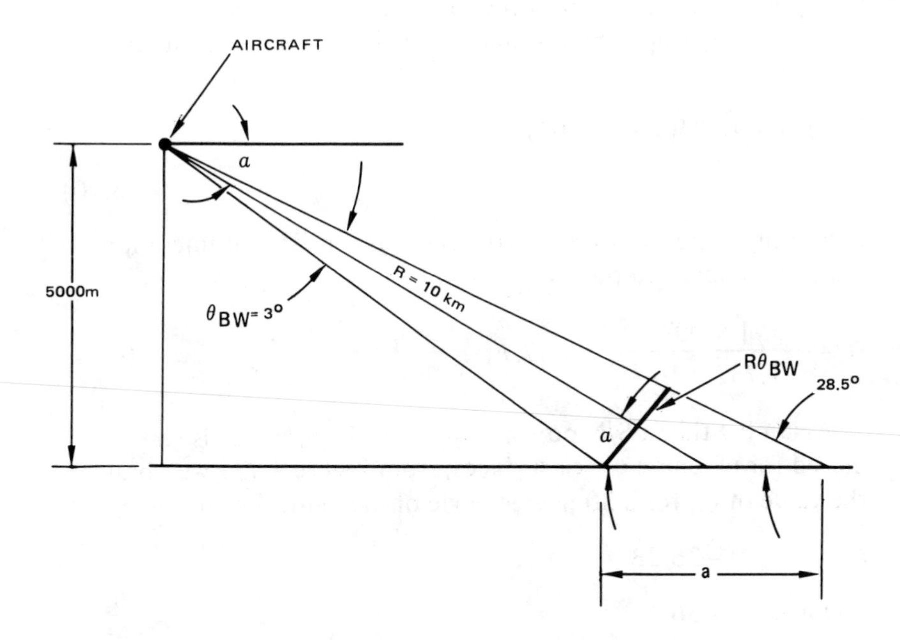

Figure 6-8 Aircraft Flight of the Design Example

where the value of α is computed from Figure 6-8 to be 30 degrees. The desired resolution of 3 meters can be achieved by a pulse duration of

$$\tau = \frac{(\Delta R_x)(2 \cos \alpha)}{c}$$

$$= \frac{(3)(2)(\cos 30°)}{3 \times 10^8}$$

$$= 1.732 \times 10^{-8} \text{ seconds} \qquad (6.8)$$

This value corresponds to a compressed pulse frequency excusion Δ of

$$\Delta = 1/\tau$$

$$= 5.77 \times 10^7 \text{ Hz} \qquad (6.9)$$

Using this frequency excusion with an uncompressed pulse of 1 μs duration, as given in the problem, we will get a pulse compression ratio of

$$D = (1 \times 10^{-6})(5.77 \times 10^{7})$$
$$= 57.7 \tag{6.10}$$

This compression ratio is also the same as the ratio of uncompressed and compressed pulses

$$D = \frac{1 \times 10^{-6}}{1.732 \times 10^{-8}} = 57.7 \tag{6.11}$$

Assuming that a S/N power ratio of 3 dB (S/N = 2) is required for mapping of sea surfaces, from Figure 3-13a we obtain the value of σ_o for a 30 degree angle of incident. From this figure

$$\sigma_o = -36 \text{ dB}$$
$$10 \log \sigma_o = -36$$
$$\sigma_o = 10^{-3.6}$$
$$= 2.51 \times 10^{-4} \tag{6.12}$$

With this information we can compute the average power requirements of the radar from (3.41). Thus the required average power will be

$$\overline{P} = \frac{(4\pi)^3 \, R^3 \, kTL}{G^2 \, \lambda^2 \, \sigma_o \, \Delta R_x} \left(\frac{2V}{\lambda}\right)\left(\frac{S}{N}\right) \sin\theta \tag{6.13}$$

Using appropriate parameters in the above equation we get

$$\overline{P} = \frac{(4\pi)^3 (10^4)^3 (1.38 \times 10^{-23})(600)(10)}{(3600)^2 (0.03)^2 (2.51 \times 10^{-4})(3)} \text{ X}$$
$$\left(\frac{2 \times 250}{0.03}\right) (2) \sin 90^\circ$$
$$= 0.623 \text{ watt .} \tag{6.14}$$

The peak power can be computed from (3.30) as

$$P = \frac{\overline{PW}}{f_r}$$

$$= \frac{(0.623)(10^6)}{1500}$$

$$= 415 \text{ watts} \qquad (6.15)$$

The doppler filter bandwidth for a $\frac{D}{2}$ resolution of 3 meters can be computed from (2.13) as

$$\Delta f_i = \frac{VD \sin \theta}{\lambda R}$$

$$= \frac{(250)(6)}{(0.03)(10000)}$$

$$= 5 \text{ Hz} \qquad (6.16)$$

Using this filter bandwidth, the corresponding filter time constant will be

$$T = 1/\Delta f_i = 0.2 \text{ second} \qquad (6.17)$$

This will result in a synthetic array length from (2.15) of

$$L = VT = (250)(0.20) = 50 \text{ meters} \qquad (6.18)$$

Thus, the data must be collected for an equivalent synthetic array length or a flight length of 50 meters to satisfy filter time constant.

The number of filters for each range gate now becomes

$$m_r = \frac{50}{3} = 16.6 \cong 17 \,, \qquad (6.19)$$

where 3 is the azimuth resolution value.

To calculate the number of range gates we compute a, the coverage of one beamwidth on the ground from Figure 6-8

$$a = R\theta_{BW}/\sin\alpha$$

$$= \frac{(10000)(3/57.3)}{\sin 30°}$$

$$= 1047 \text{ meters} \qquad (6.20)$$

Using this and an x-direction resolution of 3 meters, the number of range gates becomes

$$N_R = \frac{1047}{3} = 349 \cong 350 \qquad (6.21)$$

Using the number of filters per range gate from (6.19) we can compute the total number of filters as

$$m = (350)(17) = 5950 \qquad (6.22)$$

The limits of the PRF can also be checked using equation (3.20)

$$\frac{c}{2R_{max}} \geqslant PRF \geqslant \frac{2V}{\ell}$$

$$\frac{3 \times 10^8}{2 \times 5000 \csc 28.5} \geqslant 1500 \geqslant \frac{2 \times 250}{0.573}$$

$$14{,}314 \geqslant 1500 \geqslant 872\ , \qquad (6.23)$$

where 28.5 is the incident angle to R_{max} as shown in Figure 6.8. The selected PRF of 1500 fits the limits of equation (6.23).

Digital signal processing requirements of the example problem can be computed from the given equations. The number of samples per range gate N_D can be computed from equation (5.35) as

$$N_D = \frac{\lambda R}{(\delta AZ)^2}$$

$$= \frac{(0.03)(10000)}{(3)^2}$$

$$= 33.33 \text{ samples} \qquad (6.24)$$

This value can also be obtained by using the number of filters multiplied by the filter bandwidth for total frequency excusion. This number is then multiplied by 2, which is the number of samples per cycle, and also by the duration of sampling time

$$N_D = 17 \times 5 \times 2 \times 0.2 = 34 \text{ samples} \tag{6.25}$$

The discrepancy between (6.24) and (6.25) is due to approximations used in obtaining the parameters of (6.25). Selecting the higher binary number to N_D value we get

$$N_D = 64 = 2^6 \tag{6.26}$$

The bulk memory size will then be given by (5.35) as

$$N = (64)(350) = 2.24 \times 10^4 \text{ words} \tag{6.27}$$

Calculation rate, as the number of complex multiplications per second, is given by (5.39) as

$$\begin{aligned} NTR &= \frac{V N_R N_D \log_2 N_D}{\delta AZ} \\ &= \frac{(250)(350)(64)(6)}{3} \\ &= 1.12 \times 10^7 \text{ multiplications/second} \end{aligned} \tag{6.28}$$

Assuming that it is required to handle a S/N power ratio of no more than 40 dB, according to equation (5.40) the bulk memory and the arithmetic unit of the computer should operate with a word length consisting of eight binary bits.

The number of samples of 64 and the time duration of 0.20 second will result, equation (5.36), in the number of samples per second of

$$f_s = (64)/0.2 = 320 \text{ samples/second} \tag{6.29}$$

Since the PRF of 1500 Hz is considerably larger than f_s, the input signal should be prefiltered and sampled at a rate of 320 samples per second as given by (6.29).

6.3 REAL TIME SAR IMAGE PROCESSING EXAMPLE

This example of SAR signal processing is obtained from a paper by Wayne E. Arens, and published in the Proceedings of the Synthetic Aperture Radar Technology Conference, March 8-9-10, 1978, Las Cruces, New Mexico. The paper appears in Section V of this publication.

It is desired to develop signal processing requirements for a 200-meter resolution synthetic array radar system to be placed in an orbiting satellite. The geometry of the orbiting satellite is given in Figure 6-9. Radar and space parameters are specified in Table 6-1.* As an aid to computations, Table 6-4, given at the end of this chapter, has been prepared. This table includes all of the important SAR formulas which are derived in this book.

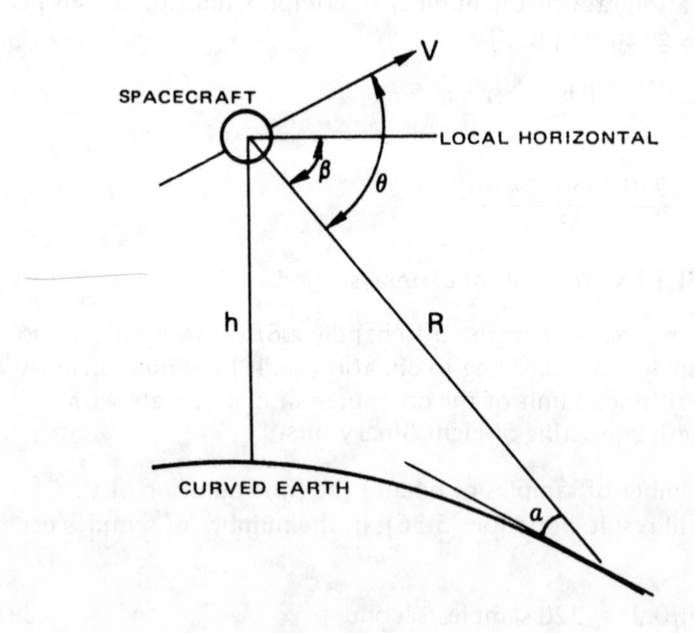

Figure 6-9 Spacecraft Geometry of Example Problem

*Page 138.

6.3.1 Radar System Design Parameters

Referring to Table 6-1 the transmit frequency of 1275 MHz will result in a wavelength λ of

$$\lambda = \frac{c}{f}$$

$$= \frac{3 \times 10^8}{1275 \times 10^6} \tag{6.30}$$

$$= 0.235 \text{ meter}$$

The pulse repetition frequency (PRF) of 1225 Hz will result in an interpulse period (IPP) of

$$IPP = \frac{1}{PRF}$$

$$= 816\ \mu \text{ seconds} \tag{6.31}$$

Spacecraft velocity, altitude, squint angle, look angle, angle of incidence and range are as given in Table 6-1. Antenna width ℓ_1 of 2 meters will correspond to a beamwidth of

$$\Omega_r = \frac{\lambda}{\ell_1}$$

$$= 0.1175 \text{ radian}$$

$$= 6.73 \text{ degrees} \tag{6.32}$$

as given in the table. Using this, the swath width, or the range coverage, of the 3-dB beamwidth on the ground can be computed as

$$SW = \frac{R\lambda}{\ell_1 \sin \alpha}$$

$$= \frac{(559)(0.235)}{(2) \sin 41°}$$

$$= 100 \text{ km} \tag{6.33}$$

Range pulse width T, together with pulse compression frequency excursion Δ, are given as 33.8 μsec and 1.25 MHz, respectively.

Antenna length ℓ_2 will result in an azimuth beamwidth of 1.12 degrees as given. The azimuth swath length will be

$$SL = \frac{R\lambda}{\ell_2}$$

$$= \frac{(559)(0.235)}{12}$$

$$= 10.94 \text{ km} \quad (6.34)$$

Using this and the spacecraft velocity of 7.14 km/s we obtain the time to fly out one beamwidth T_b as

$$T_b = \frac{SL}{V}$$

$$= \frac{10.94}{7.14}$$

$$= 1.53 \text{ seconds} \quad (6.35)$$

The total frequency spread or azimuth bandwidth can be computed as follows

$$\Delta f_d = \frac{2V}{\ell_2}$$

$$= \frac{2 \times 7140}{12}$$

$$= 1.19 \text{ kHz} \quad (6.36)$$

The oversample factor of 1.2 specified in Table 6-1 will be used in sampling rate calculations.

6.3.2 Signal Processing Parameters

Using SAR equations derived in this book and summarized in Table 6-4 given at the end of this chapter, we can derive the radar signal processing values given in Table 6-2.*

Pulse compression ratio is obtained from

$$\begin{aligned} D &= T\Delta \\ &= (33.8 \times 10^{-6})(1.25 \times 10^{6}) \\ &= 42.25 \end{aligned} \tag{6.37}$$

Range resolution without pulse compression is given as

$$\begin{aligned} \Delta R_X &= \frac{c\tau}{2 \cos a} \\ &= \frac{3 \times 10^{8} \times 33.8 \times 10^{-6}}{2 \cos 41^\circ} \\ &= 6718 \text{ meters} \end{aligned} \tag{6.38}$$

Using the pulse compression value of 42.25, the range resolution becomes

$$\begin{aligned} \Delta R_X &= \frac{6718}{42.25} \\ &= 159 \text{ meters} \end{aligned} \tag{6.39}$$

The oversampling factor of 1.2 will change this value to

$$\begin{aligned} \Delta R_X &= 159 \times 1.2 \\ &= 191 \text{ meters} \end{aligned} \tag{6.40}$$

The sampling rate of the pulse compression filter will equal twice its frequency excursion. This value is further multiplied by the oversampling factor, resulting

*Page 139.

$$\begin{aligned} SR &= (2)\,(\Delta)\,(f_{os}) \\ &= 2 \times 1.25 \times 1.2 \\ &= 3 \text{ MSPS}, \end{aligned} \tag{6.41}$$

where MSPS represents megasamples per second.

The sampling time duration for range computation with pulse compression can be obtained using the following considerations. The swath width of 100 km with a range resolution value of 191 meters will result in

$$\begin{aligned} N_R &= \frac{100 \times 10^3}{191} \\ &= 523.5 \\ &\cong 524, \end{aligned} \tag{6.42}$$

where N_R is the number of range gates or range bins. Each range bin is time equivalent to a pulse width derived using pulse compression ratio D and uncompressed pulse duration

$$\begin{aligned} \tau &= \frac{T}{D} \\ &= \frac{33.8}{42.25} \\ &= 0.8\ \mu\text{sec} \end{aligned} \tag{6.43}$$

Using this, the time equivalent of 524 range bins becomes

$$\begin{aligned} T_s &= (524)\,(0.8) \\ &= 420\ \mu\text{sec} \end{aligned} \tag{6.44}$$

Thus, for each interpulse period of 816 μsec, only 420 μsec worth of data need to be sampled for pulse compression range determination. This window is selected through a knowledge of radar-target range and other spacecraft characteristics. The number of samples during this sampling period will be

$$N_s = (420 \times 10^{-6})\,(3 \times 10^{6})$$

$$= 1260 \text{ samples,} \qquad (6.45)$$

where N_s represents the number of samples.

For an azimuth resolution of D/2 equal to 200 meters (D=400 meters) the filter frequency bandwidth will be (note that D is also used for pulse compression ratio)

$$\Delta f_i = \frac{D\,V \sin\theta}{\lambda\,R}$$

$$= \frac{(400)(7.140)(\sin 90°)}{(0.235)(559)}$$

$$= 21.74 \text{ Hz}$$

$$\cong 22 \qquad (6.46)$$

This will result in a filter time constant or coherent integration time of

$$T = \frac{1\,(1.2)}{\Delta f_i}$$

$$= 0.0545 \text{ second} \qquad (6.47)$$

Note the inclusion of oversampling factor $f_{os} = 1.2$ in the coherent integration time calculation.

The number of pulses coherently integrated now becomes

$$N_P = T(PRF)$$

$$= (0.0545)\,(1225)$$

$$= 67 \text{ pulses} \qquad (6.48)$$

The synthetic array length can be computed using coherent integration time and spacecraft velocity

$$L = VT$$

$$= (7140)(0.0545)$$

$$= 389.1 \text{ meters} \tag{6.49}$$

Using an azimuth resolution of 200 meters, this will result in

$$N_f = \frac{L}{D/2}$$

$$= \frac{389.1}{200}$$

$$= 2, \tag{6.50}$$

where N_f is the number of filters per range gate for the coherent integration time.

The total number of doppler filters in an azimuth swath can be computed by

$$N_d = \frac{SL}{D/2}$$

$$= \frac{10.94 \times 10^3}{200}$$

$$= 55 \text{ filters} \tag{6.51}$$

The number of independent looks can be obtained by dividing the number of azimuth filters by the number of filters per range gate, i.e.,

$$\text{Looks} = \frac{55}{2}$$

$$\cong 28 \tag{6.52}$$

Another method of obtaining this value is by considering time in beam and the coherent integration time, i.e.,

$$\text{Looks} = \frac{1.53}{0.0545}$$

$$= 28 \tag{6.53}$$

This means that the antenna beam coverage is sufficient to result in 28 independent returns from each ground resolution cell. This mechanization will, of course, require 55 azimuth filters per range gate, instead of only two required for synthetic flight length L. The return from these looks can be integrated in a multi-look processor.

6.3.3 Power Requirements

It is desired to obtain power requirements of the radar using the following assumptions.

Backscattering coefficient σ_o	$= 10^{-3}$
Required signal to noise ratio (S/N) for detection	= 3 dB = 2
Equivalent noise temperature, T	= 1000 °K
System losses, L	= 10 dB = 10
Antenna efficiency, η_a	= 0.70 (6.54)

The antenna gain can be computed using antenna width and length data of Table 6-1 and the following equation

$$G = \frac{4\pi A}{\lambda^2} \eta_a$$

$$= \frac{4\pi \ell_1 \ell_2}{\lambda^2} \eta_a , \tag{6.55}$$

where η_a is the antenna efficiency. Or

$$G = \frac{(4\pi)(2)(12)}{(0.235)^2} (0.70)$$

$$= 1216 \tag{6.56}$$

Using the average power equation

$$\overline{P} = \frac{(4\pi)^3 R^3 kTL}{G^2 \lambda^2 \sigma_o (\Delta R_x)} \left(\frac{2V}{\lambda}\right)\left(\frac{S}{N}\right) \sin\theta\,, \tag{6.57}$$

we get

$$P = \frac{(4\pi)^3 (559 \times 10^3)^3 (1.38 \times 10^{-23})}{(1216)^2 (0.235)^2 (10^{-3})(191)(0.235)} \times$$

$$(1000)(10)(2 \times 7140)(2)(1)$$

$$= 372 \text{ watts} \tag{6.58}$$

Using this value, the peak power requirements will be

$$P = \frac{\overline{P}}{\tau f_r}$$

$$= \frac{372}{33.8 \times 10^{-6} \times 1225}$$

$$= 8.98 \text{ kw}, \tag{6.59}$$

where P represents peak power, τ the pulse width, and f_r pulse repetition frequency. Note that this power requirement does not take into consideration the multi-look provisions and the resulting S/N gain due to integration of returned energy from many "looks." The integration of energy from 28 looks will result in a theoretical 28-fold increase in signal to noise ratio. This in turn will reduce power requirements by a factor of 28, reducing average power required to 13.2 watts and peak power required to 321 watts as given in Table 6-2.

6.3.4 Signal Processing Mechanization

In addition to the radar inputs, SAR mechanization will require spacecraft attitude, position, and also antenna beam pointing inputs. This information is provided to the motion compensation computer through various spacecraft sensors. Table 6-3* gives a summary of accuracies with which these values may be provided. This table also includes the effects of inaccuracies on various SAR outputs. Basically the angular inaccuracies will reduce S/N values as they will change the position of the antenna beam, reducing, in effect, the antenna gain illuminating a given area. Position inaccuracies will degrade resolution, as resolution values in x- and y-directions depend, respectively, on measured range and spacecraft velocity. The accuracy values of Table 6-3 were based on a nominal resolution of 200 meters ±25 percent.

Using the data of Tables 6-2 and 6-3, Figure 6-10 gives a possible SAR processor block diagram. Referring to this figure, the data interface unit receives the radar data and spacecraft parameter data required to process SAR images. The range correlator performs the range correlation function. The azimuth correlator performs the azimuth correlation function and the antenna beam pointing

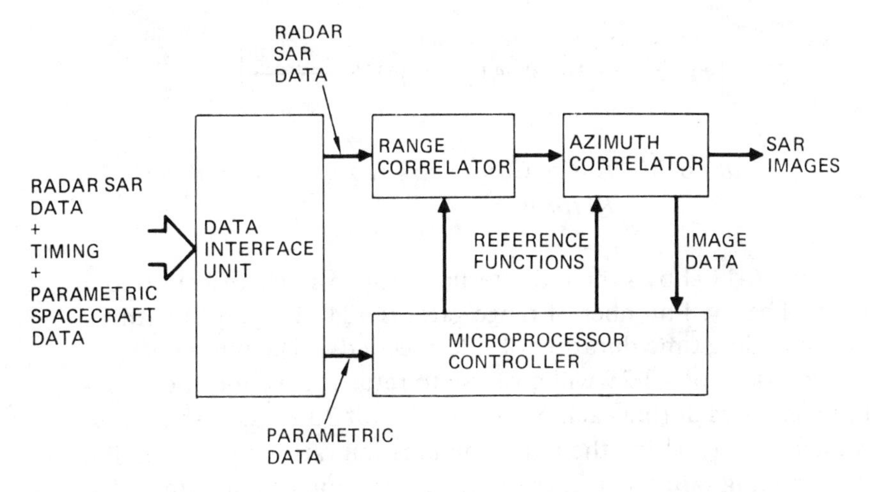

Figure 6-10 SAR Signal Processor Block Diagram

*Page 140.

correction. The microprocessor controller computes the necessary corrections and effects control functions for all of the elements of the processor.

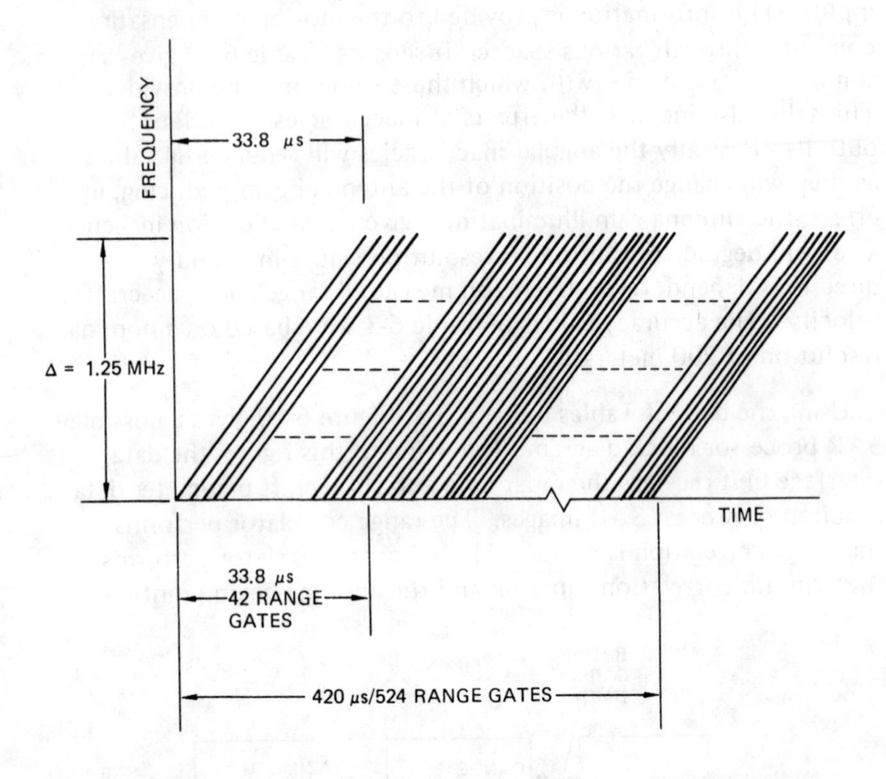

Figure 6-11 Range Gates and Frequency Modulated Returns

Figure 6-11 shows the radar returns from a single transmitted pulse. The total number of range gates is 524 (Table 6-2) with a corresponding time duration of 420 μseconds. The pulse compression ratio of 42.25 will give rise to returns covering about 42 range gates during each pulse duration of 33.8 μs, as shown in Figure 6.11. Using the pulse compression ratio of 42.25 and oversampling value of 1.2, the range correlation can be affected by a 50 stage (42.25 x 1.2 = 50) transversal filter as shown in

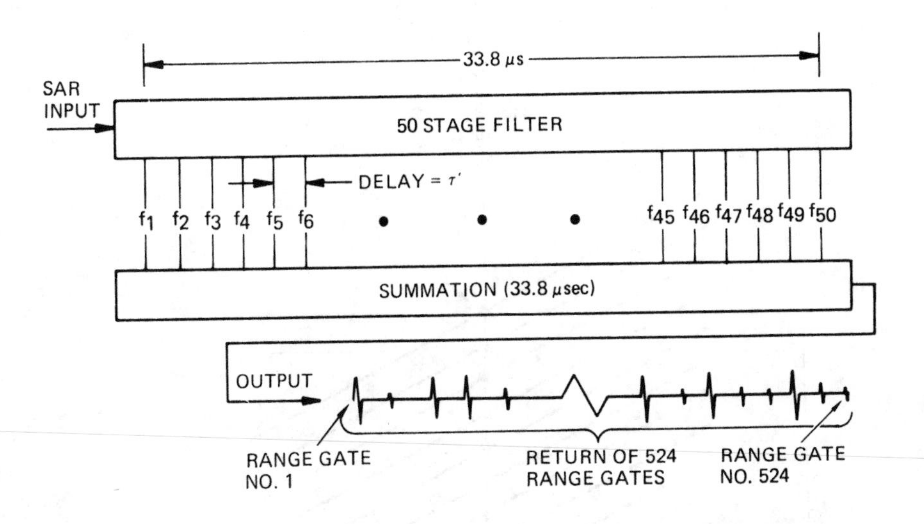

Figure 6-12 Range Correlator Consisting of Transverse Filter

Figure 6-12.* The SAR input data of the filter passes through 50 stages of timed frequencies $f_1 \rightarrow f_{50}$ with time delay between each stage of 33.8/50 = 0.68 μs. The frequency excursion of $\Delta = 1.25$ MHz is, in effect, divided between 50 frequency increments, each with a separation of $\Delta/50$ MHz. After summation the output will consist of 524 pulses spread over a time duration of 420 μs where each pulse corresponds to one range gate. The summation operation of Figure 6-12 is done in a manner similar to that of the Barker digital code calculations shown in Figure 5-10 of Chapter 5. The frequency contents of each range gated pulse will include the doppler shift information needed for azimuth correlation.

*S. C. Iglehart, "Some Results on Digital Chirp", *IEEE Transaction on Aerospace and Electronic Systems,* January 1978. Pages 118-127.

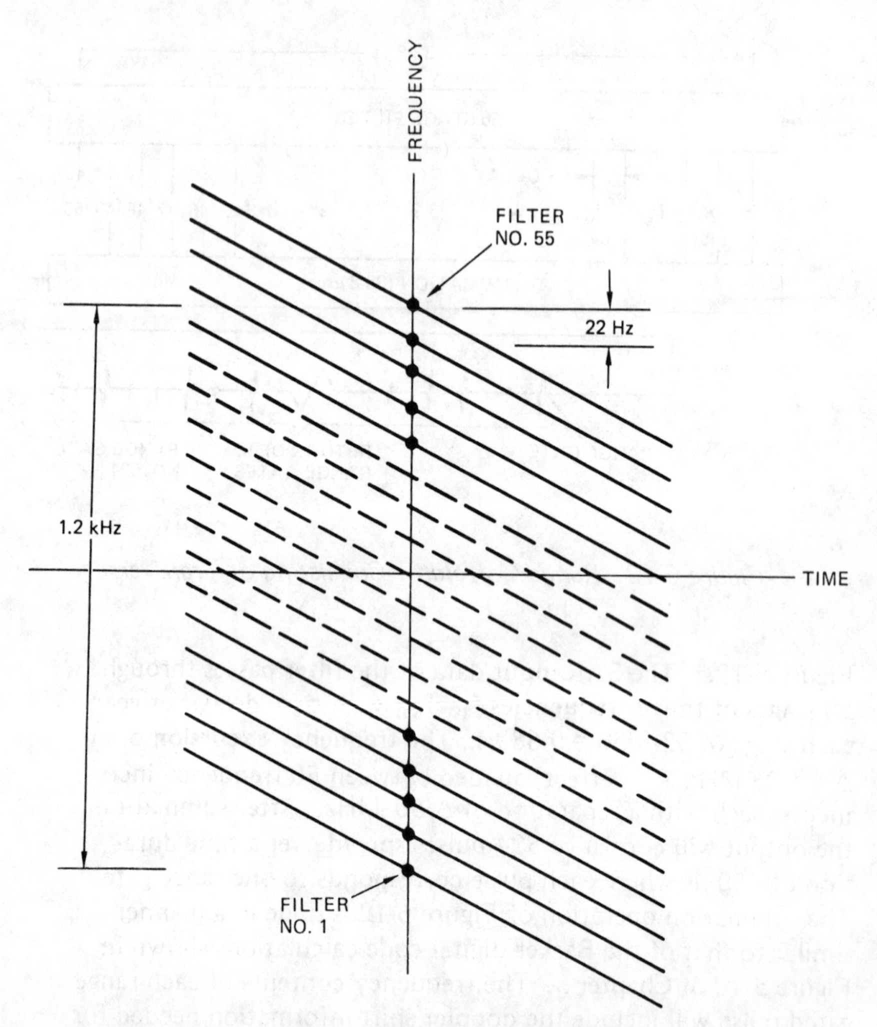

Figure 6-13 Frequency-Time Relationship of Azimuth Returns

Figure 6-13 gives a graphic illustration of return for a single range gate plotted on a frequency time diagram. Each return will have a frequency spacing of 22 Hz with 55 doppler filters formed for a total frequency excursion of 22 x 55 = 1.21 kHz. The corresponding value given as azimuth bandwidth in Table 6.1 is 1.19 kHz. Each slanted frequency line represents the return from many pulses, although it is given as a continuous return.

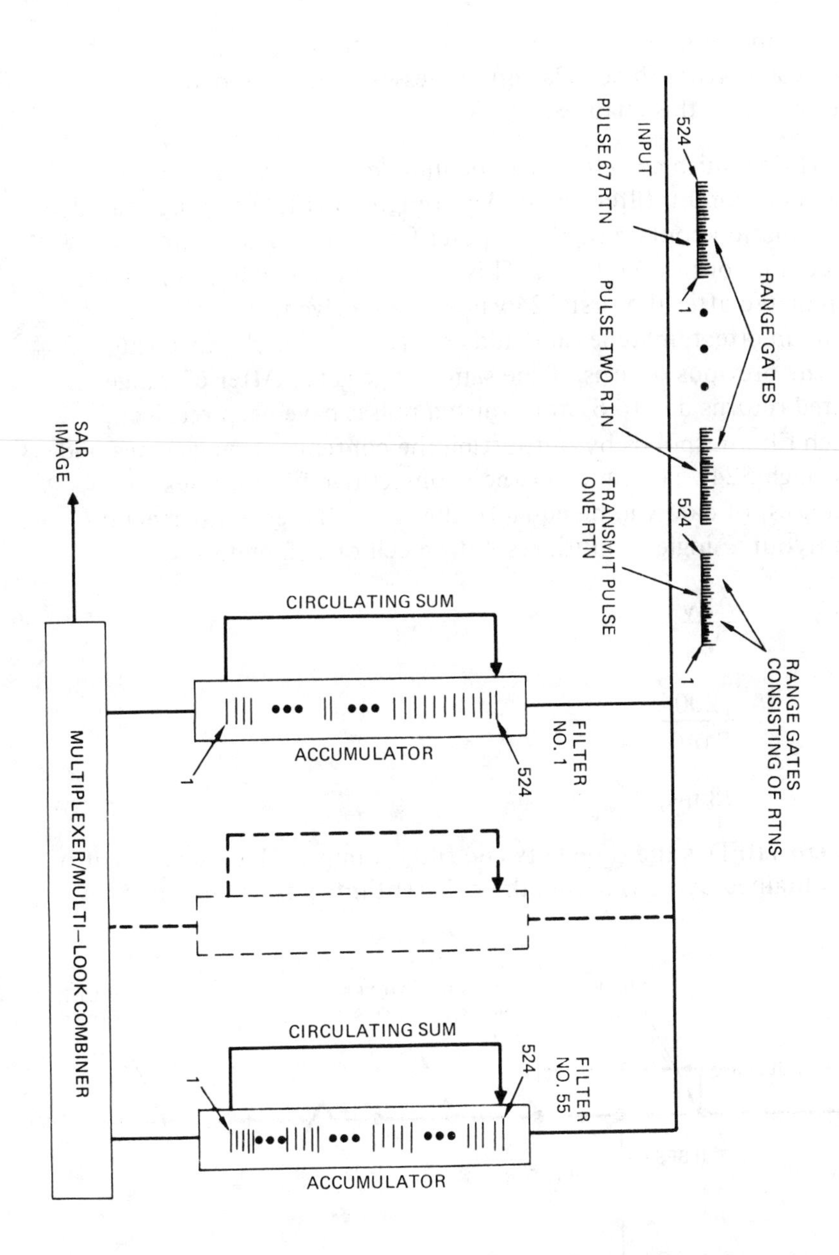

Figure 6-14 Set of 55 Parallel Azimuth Filters for Azimuth Correlation

Comparison of Figures 6-11 and 6-13 shows the similarity of range and azimuth correlation processes, as discussed at the beginning of this chapter.

The azimuth correlator can be implemented by a set of 55 parallel azimuth filters as shown in Figure 6-14. The range gated information containing the doppler frequency values enters the accumulator of each filter. This information is time-stepped and circulated after the first 524 values are received. The circulating sum, in effect, matches and adds the return of each range gate to the previous returns of the same range gate. After 67 range gated returns due to 67 transmitted pulses have been received, each filter responds by outputting the contents of range gates one through 524. The time between consecutive filter dumps (responses) of data will be equal to the time it takes the spacecraft to flyout a single azimuth resolution cell of 200 meters, or

$$\begin{aligned} \text{TBFD} &= \frac{\Delta R_y}{V} \\ &= \frac{200}{7140} \\ &= 28 \text{ ms}, \end{aligned} \tag{6.60}$$

where TBFD is the time between filter dumps. This value can also be obtained by considering the coherent integration time of 54.5

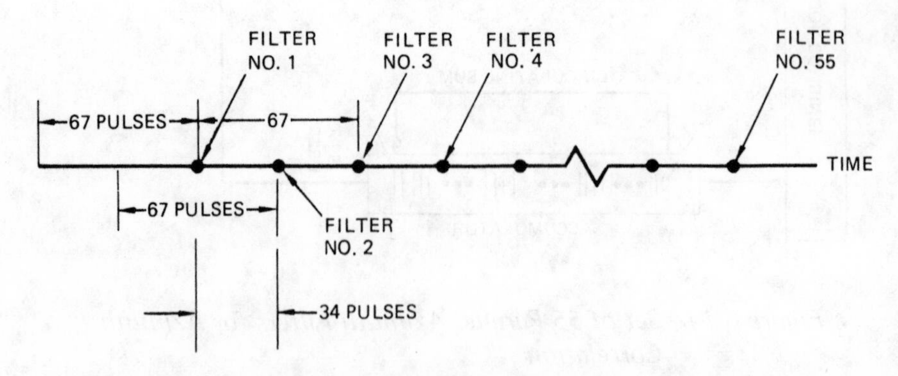

Figure 6-15 Filter Data Dumps

ms and the fact that there are two filters per range gate for the synthetic array duration L, i.e., 54.5/2 = 28 ms. The corresponding number of returned pulses will be

$$\begin{aligned} PBFD &= (TBFD)\,(PRF) \\ &= (0.028)\,(1225) \\ &= 34.3 \text{ pulses,} \end{aligned} \tag{6.61}$$

where PBFD is pulses between filter dumps. Figure 6-15 shows the start of the filter response process. The multiplexor unit of Figure 6-14 acts as an interface for sorting the data.*

Thus, as the satellite travels every 28 ms, the SAR maps out a region consisting of 524 range gates (524 x 191) = 10 km by one azimuth filter, or 200 meters.

Following azimuth correlation and a flight duration of (0.028) (55) = 1.54 seconds, a SAR image consisting of integration of S/N from 28 independent looks can be made. This will result, following the multiplexer and multilook combiner stage of Figure 6-14, in a mozaic of 524 range gates by 55 filters or 100 km by 10.9 km.

---

*Note that the implementation of the azimuth correlator would be easier if the number of filter stages and the number of accumulated pulses were equal, i.e., $N_p = N_2 = 60$. This will result in PBFD = 30.0. In this case, the resolution may not be equal to 200 meters, as specified in the statement of the problem.

TABLE 6-1 RADAR SYSTEM DESIGN PARAMETERS

PARAMETER	SYMBOL	VALUE
Transmitter Frequency	f	1275 MHz
Wavelength	λ	0.235 m
Pulse Repetition Frequency	PRF	1225 Hz
Interpulse Period	IPP	816 μsec
Spacecraft Velocity	V	7.14 km/s
Altitude	h	375 km
Squint Angle	θ	90°
Look Angle	β	46°
Incident Angle	α	41°
Range	R	559 km
Antenna Width	ℓ_1	2 m
Range Beamwidth	Ω_r	6.73°
Swath Width (Range)	SW	100 km
Range Pulsewidth	T	33.8 μsec
Range Bandwidth	Δ	1.25 MHz
Antenna Length	ℓ_2	12 m
Azimuth Beamwidth	Ω_a	1.12°
Swath Length (Azimuth)	SL	10.9 km
Time in Beam	T_b	1.53 sec
Azimuth Bandwidth	Δf_d	1.19 kHz
Over Sample Factor	f_{os}	1.2

TABLE 6-2 SIGNAL PROCESSING PARAMETERS

PARAMETER	SYMBOL	VALUE
Pulse Compression Ratio*	D	42.25
Range Resolution	ΔR_x	191 m
Range Sampling Rate	SR	3 MSPS
Number of Range Gates	N_R	524
Sampling Time Duration	T_s	420 μsec
Number of Samples	N_s	1260
Azimuth Resolution	D/2	200 m
Azimuth Filter Bandwidth	Δf_i	22 Hz
Coherent Integration Time	T	0.0545 sec
Number of Pulses Integrated	N_P	67
Number of Doppler Filters	N_d	55
Required S/N for Detection	S/N	3 dB
Equivalent Noise Temperature	T	1000° K
System Losses	L	10 dB
Antenna Efficiency	η_a	0.70
Power Required		
Average	$\overline{P}$	13.2 w
Peak	P	321 w

*Note that D is also used for azimuth resolution.

TABLE 6-3 SPACECRAFT SUPPLIED PARAMETERS

PARAMETER	ACCURACY	EFFECT OF INACCURACY
Spacecraft Attitude		
Roll	±1°	Reduce S/N
Pitch	±0.4°	Reduced S/N and
Heading	±0.4°	Doppler Ambiguity
Spacecraft Motion*		
Position	±1 km	Resolution
Velocity	±60 m/s	Degradation
Acceleration	±0.1 m/s^2	
Antenna Position		
Azimuth	±0.4°	Reduced S/N,
Elevation	±0.4°	Doppler Ambiguity

*The accuracy values are for each one of x y z directions.

6.4 SAR EQUATIONS AND PARAMETERS

The following table lists all of the SAR equations and parameters derived in this book. Parameters are defined as close to equations as possible. Because of the great number of equations involved, in a number of cases the same symbol is used for two different parameters. To eliminate confusion, there may be some redundancy in defining parameters and symbols of the table.

TABLE 6-4 SAR EQUATIONS AND PARAMETERS

NUMBER	EQUATIONS	PARAMETERS	
1	$\lambda = \frac{c}{f}$	λ	Wavelength
		c	Velocity of Propagation
		f	Frequency
2	$\Delta R_x = \frac{c\tau}{2 \cos \alpha}$	ΔR_x	Resolution in the X-direction or Perpendicular to Flight Path
		τ	Pulse Duration
		α	Angle of Incidence, Measured from Horizontal Flight Path
3	$\gamma = \frac{\lambda}{\ell}$	γ	Antenna Halfpower Beamwidth in Radians
		λ	Wavelengths
		ℓ	Antenna Dimension
4	$G = \frac{4\pi A}{\lambda^2}$	G	Antenna Boresight Gain
		A	Antenna Aperture Area
		λ	Wavelength
5	$SW = \frac{R\lambda}{\ell \sin \alpha}$	SW	Swath Width in Range
		R	Range to Target
		α	Angle of Incident
6	$SL = \frac{R\lambda}{\ell}$	SL	Swath Length in Azimuth
		R	Range to Target
7	$T_b = \frac{SL}{V}$	T_b	Time to Fly Out One Antenna Beamwidth
		V	Velocity

NUMBER	EQUATIONS	PARAMETERS	
8	$f_d = -\frac{2\dot{R}}{\lambda}$	f_d $\dot{R}$	Doppler Shift Range Rate
9	$\frac{D}{2} = \frac{\lambda R \Delta f_i}{2\,V \sin\theta}$ or $\Delta f_i = \frac{DV \sin\theta}{\lambda R}$	$\frac{D}{2}$ Δf_i V θ	Resolution in Azimuth, Y Direction or Along the Flight Path Frequency Equivalent of Resolution $\frac{D}{2}$. Flight Velocity Squint Angle or Angle Between V and R
10	$T = \frac{1}{\Delta f_i}$ $= \frac{\lambda R}{DV \sin\theta}$	T	Matched Filter Time Constant or Coherent Integration Time
11	$L = VT$ $= \frac{\lambda R}{D \sin\theta}$	L T V	Equivalent Synthetic Array Length Matched Filter Time Constant Flight Velocity
12	$\frac{D}{2} = \frac{\lambda R}{2L \sin\theta}$	$\frac{D}{2}$	Resolution Along the Flight Path in Terms of Equivalent Antenna Length
13	$\frac{3D}{2}$		Practical Attainable Resolution (Azimuth)
14	$\frac{\ell}{2}$	 ℓ	Maximum Attainable Resolution (Focused) Antenna Length (Actual)

NUMBER	EQUATIONS	PARAMETERS	
15	$\frac{D}{2} = \frac{\sqrt{\lambda R}}{2}$		Attainable Resolution in Y-Direction for an Unfocussed Array
16	$\phi = \frac{2\pi V^2 t^2}{\lambda R}$	ϕ	Two-Way Phase Shift in Radians
		t	Time
17	$f = \frac{2V^2 t}{\lambda R}$	f	Two-Sided Frequency Excursion
18	$\Delta f_d = \frac{2V}{\ell}$	Δf_d	Total SAR Frequency Spread in Azimuth
	$= \frac{2V}{D}$	V	Velocity
		ℓ	Antenna Length (Width)
		$D/2$	Resolution in Azimuth
19	$N_R = \frac{R_{max} - R_{min}}{d_r}$	N_R	Number of Range Gates
	$= \frac{R\lambda}{\ell(\tan\alpha)d_r}$	R_{max}	Maximum (3-dB) Antenna Illumination Range
	where	R_{min}	Minimum (3-dB) Antenna Illumination Range
	$R \cong (R_{max} + R_{min})/2$	d_r	Range Resolution Along $R(d_r = c\tau/2)$
		ℓ	Antenna Length
		α	Angle of Incidence
		R	Average Range
20	$N_D = (2)(\Delta f_d)(T)$	N_D	Number of Samples for Azimuth Spectrum Analysis per Range Gate
	$= \frac{4\lambda R}{D^2}$	Δf_d	Total SAR Frequency Spread in Azimuth
		T	Matched Filter Time Constant
		$D/2$	Resolution in Azimuth

NUMBER	EQUATIONS	PARAMETERS	
21	$N = N_R N_D$	N	Total Number of Samples for N_R Range Gates
22	$\frac{2V}{\ell} \leqslant PRF \leqslant \frac{c}{2R_{max}}$		Limits of PRF Selection
		ℓ	Antenna Length (Actual)
		R_{max}	Maximum 3-dB Radar-Ground Range
23	$\left(\frac{S}{N}\right)_P = \frac{PG^2 \lambda^2 \sigma}{(4\pi)^3 R^4 (kTW) L}$	$\left(\frac{S}{N}\right)_P$	Singnal to Noise Ratio per Pulse
		S	Signal Power
		N	Noise Power (= kTW)
		P	Peak Power
		G	Antenna Gain
		λ	Transmit Wavelength
		σ	Target Cross Section
		R	Radar Target Range
		k	Boltzman's Constant
		T	Equivalent Receiver Temperature
		W	Receiver Noise Bandwidth
		L	System Losses
24	$\sigma = \left(\frac{D}{2}\right)\left(\frac{c\tau}{2\cos\alpha}\right)\sigma_o$	σ	Target Cross Section of Resolvable Area Consisting of the Product of Azimuth and Range Resolution Elements and Also the Backscattering Coefficient
		σ_o	Backscattering Coefficient

NUMBER	EQUATION	PARAMETERS	
25	$\overline{P} = \dfrac{(4\pi)^3 R^3 kTL}{G^2 \lambda^2 \sigma_o (\Delta R_x)} \left(\dfrac{2V}{\lambda}\right)\left(\dfrac{S}{N}\right) \sin\theta$	$\overline{P}$	Average Power Requirement of a Synthetic Array Radar With Coherent Integration of Returned Pulses.
		R_x	Resolution Element Along the X-Direction
		V	Flight Velocity
		$\dfrac{S}{N}$	Signal to Noise Ratio Required for Detection
		θ	Angle Between V and Antenna Pointing
26	$P = \dfrac{\overline{P}}{\tau f_r}$	P	Peak Power
		τ	Pulse Duration
		f_r	Pulse Repetition Frequency
		$\overline{P}$	Average Power
27	$D = \dfrac{T}{\tau}$ $= T\Delta$	D	Dispersion Factor of a Compressed Pulse or Pulse Compression Ratio
		T	Pulse Duration Before Compression
		τ	Pulse Duration After Compression
		Δ	Frequency Excursion of Frequency Modulated Pulse
28	$t_o = \dfrac{Tf_d}{\Delta}$	t_o	Delay Due to Target Doppler Shift f_d in a Compressed Pulse
		f_d	Doppler Frequency Shift

NUMBER	EQUATION		PARAMETERS
29	$NC = N_s \log_2 N_s$	NC	Number of Complex Multiplication Using Binary Arithmetic and the Fast Fourier Transform (FFT) for Spectrum Analysis of a Time Signal
		N_s	Number of Samples. $N_s = 2^{\gamma}$ where γ is an Integer
30	$NTR = \dfrac{VN_R N_D \log_2 N_D}{\delta AZ}$	NTR	Arithmetic Rate in Number of Complex Multiplications per Second
		N_R	Number of Range Gates
		N_D	Number of Samples for Azimuth Spectrum Analysis per Range Gate
		V	Flight Velocity
		δAZ	Azimuth Resolution $\dfrac{D}{2}$

References

1. Barton, D.K. (Editor), *Radars Volume 3, Pulse Compression,* Artech House, 1975.
2. Berkowitz, R.W. (Editor), *Modern Radar,* John Wiley, 1966, p. 227.
3. Cook, C.E. and M. Bernfeld, *Radar Signals,* Academic Press, 1967, Chapter 8.
4. Harger, R.O., *Synthetic Aperture Radar Systems,* Academic Press, 1970.
5. Hovanessian, S.A., *Radar Detection and Tracking Systems,* Artech House, 1973, Chapter 7.
6. Hovanessian, S.A., *Computational Mathematics in Engineering,* Lexington Books, 1976, Chapter 6.
7. Hovanessian, S.A., "Time Domain Analysis of Digital Signal Processing," *Computers and Electrical Engineering,* Pergamon Press, London, 1975.
8. Jensen, Homer, *et al,* "Side-Looking Airborne Radar," *Scientific American,* October 1977, pp. 84-96.
9. Kirk, John C., "Digital Synthetic Aperture Radar Technology," *IEEE International Radar Conference,* 1975, page 482.
10. Klauder, J.R. *et al,* "The Theory and Design of Chirp Radars," *Bell System Technical Journal,* Volume 39, July 1960.
11. Kock, W.E., *Engineering Applications of Laser and Holography,* Plenum Press, 1975, pp. 196-223.
12. Kovaly, J.J., *Synthetic Aperture Radars,* Artech House, 1976.
13. Preston, K., *Coherent Optical Computers,* McGraw-Hill, 1972, pp. 256-264.
14. Rihaczek, A.W., *Principles of High Resolution Radar,* McGraw-Hill, 1969.
15. Tomiyasu, K., "Review of Synthetic Aperture Radar," *Proceedings of IEEE,* May 1978, pp. 563-583.

Index

A

absorption
    atmospheric 52
    rain 56
acceleration, centripetal 25
acceleration effect 66
airborne SAR, example 71
ambiguity
    range 5, 48
    range/range rate 87
angle, squint 27
antenna
    beamwidth 6, 141
    circular 7
    efficiency 8
    fan beam 6
    gain 7, 116, 141
    pattern 6, 8
    sidelobe 8
aperture (*see* array)
Apollo Lunar Sounder 76
applications of SAR 65
arithmetic rate 146
array
    focused 23
    length 19
    radars 13
    unfocused 23
atmospheric absorption 52
attenuation of electromagnetic energy 56
average power 60, 145
azimuth
    correlator 135
    filter 135
    resolution 6, 62, 110, 142

B

backscattering coefficient 49, 53
bandwidth
    filter 16
    null-to-null 16
Barker code 90
beam antenna 6
beamwidth 6
binary bits 106
binary phase code 90
BPC (*see* binary phase code)
bulk memory 105

C

California Institute of Technology 68
cell resolution 3
centripetal acceleration 25
chirp radar 80
circular antenna 7
codes
    binary phase 90
    Barker 90
    Frank 92
    poly-phase 90
coherent
    detectors 96
    integration time 98
    phase 31, 68
complex filtering 67
compression factor 145
computer
    memory 104
    speed 104
conventional radar maps 21, 26
convolution integral 83
correlation 109
cosine weighting 82
cross-sectional area 49, 144

D

data
- dumps 136
- storage 67

design example 114, 122
detectors, coherent 96
digital codes 90, 92
digital filter 103
digital pulse compression 90
digital signal processing 96, 102, 104
discrete Fourier series 99
dispersion factor 82, 145
doppler
- filters 42
- shift 15, 85, 142

E

electronic signal processing 31, 38
equivalent antenna array 19
example radar imagery 68

F

fan beam antenna 6
fast Fourier transform 100, 114
FFT (*see* fast Fourier transform)
filter
- azimuth 135
- bandwidth 16
- data dump 136
- design 16
- doppler 42
- time constant 16
- transverse 133

filtering requirements 67
focused array 23, 27
fog absorption 56

Fourier analysis 96
    digital mechanization 102
    integral, matrix form 100
    transform 97
foldover frequency 99
Frank code 92
frequency
    excursion 143
    foldover 99
    history 38
    modulation slope 84
    phase relationship 13
    time relationship 134
Fresnel zones 32, 75

G

Gabor, Dennis 31
gain antenna 7, 141
grating plates 32
ground map 21

H

halfpower beamwidth 141
holography 31
Hughes Aircraft Company 71

I

imagery example 68
imaging radars 1, 10
infrared imagery 74
integrated sidelobe level 65
integration time, coherent 98

J

Jet Propulsion Laboratories 68

L

length, synthetic array 19, 142
low pass filter 101

M

mainlobe antenna 8
mapping
    radars 1
    resolution 3
matched filter 84
matched filter time constant 142
matched transmit/receive 4
matrix Fourier transform 100
maximum resolution 142
mechanization SAR 95
    example 131
memory computer 104
microprocessor 132
moisture content 74
molecular resonance 56
motion compensation 66, 131
multiplication, number of 146

N

number
    multiplications 146
    range gates 143
    samples 143
Nyquist sampling 44

O

oceanology 75
ocean surface backscattering 50
optical signal processing 31, 34
oversample factor 124
oxygen absorption 55

P

parallel azimuth filters 135
pattern antenna 6
peak
    power 145
    sidelobes 65

pencil beam 6
periodic function 96
perpendicular to track
resolution 4
phase
coherency 68
frequency 13
history 38
quadratic 38
shift 25, 143
phasors 102
polyphase code 90
power
average 60, 145
pulse compression 61
return 48, 61, 129
prefilter 106
PRF (*see* pulse repetition frequency)
programmable signal processor 104, 106
pulse compression 79
digital 90
doppler shift 85
factor 145
matched filter 145
power 61
S/N 62
pulse radars 3
pulse repetition frequency 45, 144

Q

quadratic phase shift 25, 38
quadrature components 100

R

radar
imaging 1, 10
mapping 21
pulse 3

stretch 114
synthetic array 13
target cross section 52
rain absorption 56
range
ambiguity 5, 48
correlator 133
gates 3, 42, 110
ground 43
range rate ambiguity 87
resolution 82, 111, 141
(*see also* x-direction resolution)
rate, arithmetic 146
Raytheon Company 11
real array 1
real time SAR 122
resolution 3, 6, 17
along flight path 142
azimuth 62, 142
cell 3
correlator 110
element 44
focused 23
frequency equivalent 142
maximum 22, 142
practical 142
range 82, 111, 141
unfocused 23, 41
resolvable area 49
resonant, molecular 56

S

sampling, Nyquist 44
SAR (*see* synthetic array radar)
Seasat-A satellite 68
sensor information 131
shift, doppler 15
sidelobes, antenna 8, 65, 67
signal to noise ratio 48, 144

signal processing 31, 34, 38, 104, 125, 131
slot noise 65
soil moisture 74
spaceborne SAR 68, 122
spherical waves 34
squint angle 27
stability, transmitter-receiver 68
stretch radars 114
swath length 141
swath width 141
synthetic array length 19, 142
synthetic array radars 3, 65, 74
mechanization 95
parameters 140

T

target cross section 49, 52, 144
time constant 16, 142
time in beam 141
transmitter-receiver stability 68
transverse filter 133
two-dimensional correlation 109

U

unfocused array 23, 27
unfocused resolution 23, 41

V

vegetation resources 75

W

water resources 75
water vapor absorption 54
wavelength formula 141
weather effects 56
weighted antenna pattern 8
weighting function 82
Westinghouse Electric Company 11

X

x-direction resolution 5, 112, 141

Y

y-direction resolution 6, 18, 20, 27, 113, 142

Z

zone plates 26, 31